T0231570

Mining
AND THE
ENVIRONMENT

INTERNATIONAL PERSPECTIVES ON PUBLIC POLICY

edited by
RODERICK G. EGGERT

John M. Olin
Distinguished Lectures
in Mineral Economics

Routledge
Taylor & Francis Group

NEW YORK AND LONDON

©1994 Resources for the Future

All rights reserved. No part of this publication may be reproduced by any means, either electronic or mechanical, without permission in writing from the publisher, except under the conditions given in the following paragraph.

Authorization to photocopy items for internal or personal use, the internal or personal use of specific clients, and for educational classroom use is granted by Resources for the Future, provided that the appropriate fee is paid directly to Copyright Clearance Center, 222 Rosewood Drive, Danvers, MA 01923, USA.

First Published 1994 by Resources for the Future

This edition published 2011 by Routledge:

Routledge
Taylor & Francis Group
711 Third Avenue
New York, NY 10017

Routledge
Taylor & Francis Group
2 Park Square, Milton Park, Abingdon,
Oxfordshire OX14 4RN

First issued in hardback 2016

Routledge is an imprint of the Taylor and Francis Group, an informa business

Library of Congress Cataloging-in-Publication Data

Mining and the environment : international perspectives on public policy / edited by
    Roderick G. Eggert.
          p.   cm.
    "John M. Olin distinguished lectures in mineral economics."
    Includes bibliographical references.
    ISBN 0–915707–72–1
    1. Mineral industries—Environmental aspects. 2. Mineral industries—
Government policy.   I. Eggert, Roderick G.
TD195.M5M576 1994
333.8'514—dc20   94-7236
                                                  CIP

This book is the product of the Energy and Natural Resources Division at Resources for the Future, Douglas R. Bohi, director. It was edited by Betsy Kulamer and designed by Brigitte Coulton. The cover was designed by Kelly Design.

ISBN 13: 978-1-138-15536-7 (hbk)
ISBN 13: 978-0-915707-72-0 (pbk)

RESOURCES FOR THE FUTURE

Directors

Darius W. Gaskins, Jr., *Chair*

Joan Z. Bernstein	Thomas J. Klutznick	Paul C. Pritchard
David R. Chittick	Frederic D. Krupp	Robert M. Solow
Anthony S. Earl	Henry R. Linden	Linda G. Stuntz
Lawrence E. Fouraker	Thomas E. Lovejoy	Linda C. Taliaferro
Robert W. Fri	Karl-Göran Mäler	Victoria J. Tschinkel
Robert H. Haveman	Laurence I. Moss	Barbara S. Uehling
Donald M. Kerr	Frank D. Press	Mason Willrich

Officers

Robert W. Fri, *President*
Paul R. Portney, *Vice President*
Edward F. Hand, *Vice President–Finance and Administration*

RESOURCES FOR THE FUTURE (RFF) is an independent nonprofit organization engaged in research and public education on natural resource and environmental issues. Its mission is to create and disseminate knowledge that helps people make better decisions about the conservation and use of their natural resources and the environment. RFF neither lobbies nor takes positions on current policy issues.

Because the work of RFF focuses on how people make use of scarce resources, its primary research discipline is economics. However, its staff also includes social scientists from other fields, ecologists, environmental health scientists, meteorologists, and engineers. Staff members pursue a wide variety of interests, including forest economics, recycling, multiple use of public lands, the costs and benefits of pollution control, endangered species, energy and national security, hazardous waste policy, climate resources, and quantitative risk assessment.

Acting on the conviction that good research and policy analysis must be put into service to be truly useful, RFF communicates its findings to government and industry officials, public interest advocacy groups, nonprofit organizations, academic researchers, and the press. It produces a range of publications and sponsors conferences, seminars, workshops, and briefings. Staff members write articles for journals, magazines, and newspapers, provide expert testimony, and serve on public and private advisory committees. The views they express are in all cases their own, and do not represent positions held by RFF, its officers, or trustees.

Established in 1952, RFF derives its operating budget in approximately equal amounts from three sources: investment income from a reserve fund, government grants, and contributions from corporations, foundations, and individuals. (Corporate support cannot be earmarked for specific research projects.) Some 45 percent of RFF's total funding is unrestricted, which provides crucial support for its foundational research and outreach and educational operations. RFF is a publicly funded organization under Section 501(c)(3) of the Internal Revenue Code, and all contributions to its work are tax deductible.

Contributors

Anthony Cox

Senior Economist, Australian Bureau of Agricultural and Resource Economics, Canberra, Australia

Roderick G. Eggert

Associate Professor, Mineral Economics Department, Colorado School of Mines, Golden, Colorado, and Editor, *Resources Policy*

David Humphreys

Deputy Chief Economist, The RTZ Corporation, London, United Kingdom, and Vice President, Euromines, Brussels, Belgium

Gustavo E. Lagos

Head of the Mining Program, Pontificia Universidad Católica de Chile, Santiago, Chile

John E. Tilton

William J. Coulter Professor and Head, Mineral Economics Department, Colorado School of Mines, Golden, Colorado, and University Fellow, Resources for the Future, Washington, DC

Alyson Warhurst

Senior Fellow, Science Policy Research Unit, University of Sussex, Brighton, United Kingdom, and Director, Mining and Environment Research Network

Contents

Preface

The time-honored saying, "You can't make an omelet without breaking eggs," translates easily into "You can't mine orebodies without insulting the environment." For many generations, this was not a problem. Instead, the problem was to find the richest ore, extract what was valuable, and go on to the next site. The miner was a semiromantic figure in the way of the Chicago stockyards or industrial smokestacks, which were glorified by Walt Whitman.

The attitude toward mining has changed in the industrialized countries and is in the process of changing in the developing ones. Government everywhere has attempted, with varying degrees of success, to lay down rules intended to keep environmentally adverse developments in check without stymieing an activity—mining—that makes an indispensable contribution to economic growth.

The hopeful view is that government and business will be sufficiently enlightened to realize that they must cooperate in the search for the common good. The skeptical view is that each side will cling to what it perceives to be in its own best interest and that the path toward reconciliation is bound to be rocky.

The five lectures presented in this volume address the current relationships between mining and government in different ways but always with the perspective of conflict and reconciliation. The lectures were originally offered at the Mineral Economics Department of the Colorado School of Mines in late 1992, under the auspices of the John M. Olin Distinguished Lectures in Mineral Economics. Each lecturer was invited to present a topic of his or her choosing and, in addition, to spend one or more days in residence on campus, giving both students and faculty added time to delve more thoroughly into the many aspects of the lecturers' chosen topics.

The reader will look in vain for comprehensiveness—the scope of the subject prevents it. Indeed, the value of this collection lies not in its comprehensiveness but rather in its variety of problems and case studies. Each contribution fits within the broad topic reflected in the title, but also illuminates a different facet of that topic. The volume thus has the potential to clarify concepts, illuminate viewpoints, and suggest policy remedies, making it of direct interest to those intimately involved in the management of mining enterprises and in its analysis.

Thanks are due to the John M. Olin Foundation for providing funding for the undertaking, to Resources for the Future for reviewing and publishing the volume, and to Roderick Eggert for his efforts as organizer, editor, and author of a most useful introduction.

Earlier publications in this series, which were also funded by the Olin Foundation and published by Resources for the Future, are *Mineral Wealth and Economic Development*, edited by John E. Tilton, and *Making National Energy Policy*, edited by me. All three publications have profited from a long-standing cooperation between the Colorado School of Mines and Resources for the Future. This program is formally designated as the Mineral Economics and Policy Program and is jointly chaired by John E. Tilton and myself.

Hans H. Landsberg
Senior Fellow Emeritus
Resources for the Future

Mining and the Environment: An Introduction and Overview

RODERICK G. EGGERT

Mining, by its nature, poses major environmental challenges. It creates large volumes of, for example, overburden, waste rock, tailings, acid mine drainage, airborne dust, and other contaminants. These by-products are deposited on land and in the air and water, in some cases harming human health, damaging property, and affecting fish and wildlife. Mining in remote and undeveloped areas may reduce the aesthetic value of natural environments. Most wastes from mining are nonhazardous, but some are hazardous.

That mining affects the environment has long been recognized. Georgius Agricola, writing in the 1500s in the book considered to be the first scientific and technical text on mining and metallurgy, noted:

> The strongest argument of the detractors [of mining] is that the fields are devastated by mining operations... They argue that the woods and groves are cut down, for there is need of an endless amount of wood for timbers, machines, and the smelting of metals. And when the woods and groves are felled, then are exterminated the beasts and birds, very many of which furnish a pleasant and agreeable food for man. Further, when the ores are washed, the water which has been used poisons the brooks and streams, and either destroys the fish or drives them away. Therefore the inhabitants of these regions, on account of the devas-

tation of their fields, woods, groves, brooks and rivers,
find great difficulty in procuring the necessaries of life,
and by reason of the destruction of the timber they are
forced to greater expense in erecting buildings.
(Agricola 1950, 8)

For centuries, denuded landscapes, fouled streams, and dirty air
were considered part of the price that had to be paid for mineral production. Even most early environmental legislation in the United
States, Europe, Japan, and other industrialized countries, dating back
to the 1960s and 1970s, was not designed with mining in mind. Rather
such legislation was aimed at large industrial sources of water and air
pollution. Many developing countries had little in the way of environmental policy.

Times have changed. The 1990s have witnessed a renewed concern
worldwide about the environment. The unifying theme has become *sustainability*: development that "meets the needs of the present without
compromising the ability of future generations to meet their own
needs" (World Commission on Environment and Development 1987, 8).
There is no consensus on exactly what is to be sustained and by what
means sustainability is to be achieved. But there is general agreement
that fairness to future generations requires that economic development
today not come at the expense of environmental damage that leaves
future generations worse off than the current generation.[1]

As part of this renewed environmental concern, mining is the focus
of greater environmental scrutiny today than in the past. In the United
States, for example, the desire for greater protection of the environment is a central concern of those advocating revision of the Mining
Law of 1872, which has governed mining on federal lands for more than
a century. In Australia and other countries, mining has come into conflict with desires for preservation. In Poland, the former East Germany,
and elsewhere, it has become apparent that mining took place in many
centrally planned economies with little regard for its environmental
consequences. Many developing countries are designing environmental policies where none existed previously, and these surely will affect
future mining activities.

---

[1]For an introduction to the issues surrounding sustainable development, see Darmstadter 1992.

As a consequence, the mining industry is understandably more concerned now than even five or ten years ago about the possible effects of environmental policy on its business activities. In 1991, a group of mining companies formed the International Council on Metals and the Environment to represent them in discussions on environmental issues. Most major mining companies have senior managers, often reporting directly to chief executive officers, in charge of corporate environmental affairs—something unheard of a decade ago. As Sir Derek Birkin, chairman of the mining company RTZ, notes: "For many years, the basic disciplines of the minerals industry have been separated along the four basic lines of geology, mining, mineral processing, and metallurgy but now a major new field has emerged—that of the environment" (Fifth Discipline 1993, 290).

Another indicator of this increased interest is the large and growing number of proceedings volumes from professional conferences on environmental management in mining (see, for example, Lootens, Greenslade, and Barker 1991; Van Syl, Koval, and Li 1992; and Yegulalp and Kim 1990). Other recent books serve as handbooks for mine managers (see, for example, Australian Mining Industry Council 1991; Hutchison and Ellison 1992; and Sengupta 1993).

Environmental aspects of mining, therefore, have become critical concerns both for mining companies, as they make investment and production decisions, and for governments, as they design and implement new environmental policies.

Concerns such as these about mining and the environment motivated the lectures collected here. Each lecturer seeks to contribute to the ongoing debate about public policy by highlighting a particular issue or policy area. With one exception, each provides the perspective of a specific country or group of countries. There is no attempt to be comprehensive. For the most part, the talks confine themselves to exploration, development, and mining; later stages of processing, such as smelting and refining, which produce raw and semifabricated metal, are not emphasized. The addresses focus largely on metal mining, although coal and industrial minerals appear several times in the discussion.

This collection contains the revised texts of invited lectures. The speakers have drawn upon their own unique perspectives and experiences, and no attempt has been made editorially to standardize terminology or denotation. Nor, given the public venue of their initial presentation, do the lecturers necessarily provide the extensive documentation that characterizes scholarly essays.

The lectures in this collection have been written to be accessible to nonspecialists. Reading them requires no more than general familiarity with economics, environmental policy, and the mineral industries. The presentations should be of interest not only to environmental economists, mineral economists, public policy analysts, and mining industry executives, but also to students and others wanting an introduction to many of the important public policy issues in the area of mining and the environment. For readers unfamiliar with the mining industry, the following section provides some necessary background information about mineral production and its environmental consequences.

Mining, the Environment, and Public Policy

Mineral production takes place in stages. Both the principal effects of mining on the environment and the important issues for public policy in this area are perhaps best introduced within the context of these production stages.

Mineral Exploration and Mine Development

Before a mineral deposit can be mined, it must be discovered and its economic and technical viability demonstrated; this is the *exploration* stage. The environmental disruption caused by exploration tends to be localized and minor. Most damage that does occur can be remediated relatively easily. During initial assessment of a region's geologic potential, explorationists rely heavily on satellite images, airborne geophysical surveys, and large-scale geologic maps to study large areas of land—hundreds or even thousands of square kilometers. Environmental impacts are essentially nil.

Explorationists then narrow the focus of their search to smaller, more promising areas, involving perhaps several hundred square kilometers. Typical activities include geologic mapping, geochemical sampling, and surface geophysical surveying, which are carried out on the ground without large-scale equipment. Although the environment is affected by these activities, the impacts are minor.

Only in the subsequent, subsurface examination of still smaller areas is there any appreciable environmental impact—from drilling, trenching (bulldozing a trench to examine near-surface rocks), and the

4

associated road building to provide access for drill rigs and bulldozers. Such impacts can be mitigated, albeit at a cost, by reclaiming drill sites and trenches and by revegetating roads. In some instances, the need for roads in remote areas has been eliminated by using helicopters to deliver drilling equipment.

For every one hundred or so mineral deposits that are discovered and evaluated in detail during exploration, fewer than ten on average will be prepared for production during the second stage of mineral production, *mine development*. During development, mining companies design and construct mining and beneficiation facilities, arrange for financing, provide for infrastructure, and develop marketing strategies, among other activities. The environmental impacts of these activities are more significant than those resulting from exploration but much less than those of mineral production itself.

Two types of public policy are critical during mineral exploration and mine development. The first type of public policy consists of land-use rules governing whether land is available for exploration and development. The second type, applicable on those lands available for mineral activities, consists of environmental rules governing permits, environmental impact assessments, and other preproduction activities and approvals that are necessary to proceed from exploration to mine development and mining—in short, the process of environmental compliance prior to mining.

Land-use rules are important because, before mining companies can undertake mineral exploration and development, they need access to prospective mineralized lands. To be sure, in situations where mineral rights are privately held, land access is gained through negotiation between interested private parties. But for most lands worldwide, mineral rights are held by governments.[2]

Explorationists or miners typically gain access to these lands in one of three ways: negotiation with a government agency, competitive bidding, and—in a few cases, such as in the United States—claim staking (that is, claiming the right to explore on a first-come, first-served basis

---

[2]Governments may hold mineral rights for one of two reasons. First, in most countries, mineral rights are held by the government regardless of who owns the surface rights. Second, in a relatively small number of countries, surface and subsurface rights are not separated, and mineral rights belong to whoever owns the surface. In these countries (which include the United States), the government holds significant mineral rights only when it is also the owner of significant quantities of land.

when lands are considered open for exploration unless they are specifically declared off-limits, such as for national parks or wilderness areas). Existing land-use policies have placed large tracts of land off-limits to mineral exploration and development in a number of countries, including Australia, Canada, and the United States. The desire to avoid the environmental damages of mining is an important reason behind these withdrawals of land from mineral activities.

Public policies in the second category, rules governing the preproduction process of complying with environmental rules, take a number of forms. The most common are environmental permits and environmental impact assessments. Mining companies typically are required to obtain environmental permits signifying government approval of various aspects of their mine plans, including those for reclamation, waste disposal, sewage treatment, drinking water, and construction of dams and other impoundments. Companies also often have to carry out detailed assessments of the environmental impacts of proposed mineral development, which in turn are used by governments in deciding whether to permit mine development at all.

Environmental permits and assessments are important to mining companies because they increase the time, costs, and risks associated with bringing a mine into production. Costs may rise because of expenditures on permitting and environmental assessment and on implementing changes in project design that the compliance process may require. Risks rise, from the perspective of the firm prior to mining, in the sense that governments may decide not to allow mine development after companies have spent significant sums of money on exploration.

Mining and Beneficiation

Once a mineral deposit has been discovered and developed, it is ready for the next stages in the production process: mining and beneficiation. During *mining*, metal-bearing rock called *ore* is extracted from underground or surface mines. Metal concentrations in ore vary greatly, from less than 1 percent by weight for most gold deposits to over 60 percent for some iron ore deposits; most metallic mineral deposits have ore grades in the range of 1–5 percent by weight. *Beneficiation*, sometimes called milling, usually occurs at the mine site. During this stage, ore is processed (or upgraded) into concentrates, which will be processed still further, usually in a smelter or refinery.

Mining and beneficiation can have a variety of environmental effects.[3] The most visible effect probably is the sight of land disturbed by mining and waste disposal. The environmental damage is largely aesthetic.

To put the problem of potentially unsightly land into perspective, consider the study by Johnson and Paone (1982). They estimated that over the fifty-year period 1930–1980, only 0.25 percent of the total land area of the United States was used for surface mining, disposal of wastes from surface and underground mines, and disposal of wastes from mineral beneficiation and further processing. Coal mining represented about half of this land, with mining of nonmetallic minerals accounting for about two-fifths and of metallic minerals about one-tenth. Some 47 percent of the land affected by mining and waste disposal had been reclaimed. The figures of course would vary considerably from country to country, but the essential point is that only a relatively small amount of land is involved in mining and associated waste disposal. Mining activities use much less land than agricultural production, urban development, logging and forestry, and national parks and wilderness areas.

Mining and beneficiation account for significant fractions of the total amount of solid waste generated each year. Very crude estimates compiled by the United Nations Environment Programme (1991b) indicate that mining activities, apparently in this case including oil and gas production and coal mining, account for about three-quarters of the solid wastes generated annually in Canada, one-third in the United States, one-tenth in the European Community, and one-twentieth in Japan. The substantial differences reflect differences in the size of the extractive industries relative to the total economies in these countries. More detailed data for the United States suggest that non–coal mining accounts for about one-seventh of the solid waste generated annually that is considered nonhazardous, while coal mining accounts for less than a hundredth of such wastes. Manufacturing, on the other hand, generates more than half of nonhazardous solid wastes (Office of Technology Assessment 1992).

These volumes of waste, however, are not good proxies for the amount of actual environmental damage caused by mining and benefici-

---

[3]For a more extensive overview of the environmental effects of metal mining and mineral processing, see United Nations Environment Programme (1991a).

ation. For this point to be clear, it is necessary to know more about the three important types of solid waste generated by mining and beneficiation. *Overburden* is soil and rock removed to gain access to a mineral deposit prior to surface mining. *Waste rock* is separated from ore during mining. Overburden and waste rock typically are deposited adjacent to a mine (or in a mine, in the case of waste rock from underground mining). *Tailings* are the fine waste particles that are produced during the beneficiation of ore and typically suspended in water. Tailings from surface mines usually are deposited in a tailings (or settling) pond, while those from underground mines are deposited in the mine itself. (In a few countries, tailings can be deposited directly into the environment.)

The amount of solid waste generated during mining and beneficiation essentially is determined by, first, the type of mine and, second, the ore grade, or concentration of metal in the rock that will be beneficiated. The type of mine is important because underground mines generate no overburden, and mining techniques are selective enough to extract ore with only small amounts of waste rock. Surface mines, on the other hand, usually generate more than twice as much overburden and waste rock as ore.

As an example, for underground copper, gold, silver, and uranium mines in the United States, the ratio, (overburden + waste rock):ore, is on the order of 0.1:1 to 0.3:1. For surface mines, the ratios range from 2:1 to 10:1 on average (EPA 1985; based on data from the U.S. Bureau of Mines).

Ore grade, the second determinant, governs the quantity of tailings generated by a beneficiation plant. An operation with ore grading 1 percent by weight, for example, will generate ninety-nine pounds of tailings for every pound of metal, assuming complete metal recovery. Actual recovery rates usually range between 90 and 100 percent, resulting in somewhat larger volumes of tailings.

By themselves, the solid wastes of metal mining and beneficiation would cause little environmental damage, except aesthetically. But when surface and ground waters interact with these wastes, acid mine drainage can be created, and this is probably the most serious environmental problem of metal mining and beneficiation. When water interacts in an oxidizing environment with the sulfide minerals typical of most metal mines, sulfuric acid is created. Metals then are dissolved in the resulting acidic water. Acid mine drainage can contaminate drinking water and affect aquatic and plant life if it gets into surface or ground waters.

The nature and extent of actual environmental damage caused by solid mine wastes and, in turn, acid mine drainage vary enormously from case to case, depending on several factors. The type of mineral deposit being mined is important: sulfide-poor deposits, for example, generate less of the sulfur needed to create sulfuric acid than sulfide-rich deposits, and high-grade deposits will have fewer tailings per unit of recovered metal than low-grade deposits. Mining and beneficiation techniques are important: underground mining, as noted above, creates much smaller volumes of waste per unit of metal than does surface mining, and the higher the recovery rate during beneficiation, the smaller the amount of tailings created. Climate is important: in arid regions, there is little of the water necessary to create acid mine drainage. Location and population density are important: acid mine drainage that enters streams feeding into sources of human drinking water not only destroys fish and wildlife habitats, but also damages human health. Finally, the environmental management practices of mining companies are important: waste piles that are revegetated or in some other way sealed, for example, are much less likely to be accessible to the water necessary to create sulfuric acid.

Other environmental problems may be associated with metal mining and beneficiation. Another type of water contamination is wastewater from beneficiation plants, which may contain ore material, heavy metals, thiosalts, and chemical reagents used in beneficiation. Air pollution is limited largely to airborne dust. Underground mining may lead to subsidence. (A major area not dealt with in this volume is the working environment, that is, worker health and safety; readers interested in this issue are referred to Section 11 of Hartman 1992).

The key issues for environmental policy affecting ongoing mining and beneficiation are for the most part the same as those affecting other economic activities: What should be the standards for environmental quality and how should they be determined? What policy tools—for example, direct regulation or economic incentives—are best suited for meeting these standards? How should rules be enforced?

Two aspects of mining and beneficiation noted above, however, bear on these more general questions. First, the extent to which the amount of solid waste generated from mining and beneficiation can be reduced has significant limits. Low-grade ores by their very nature are going to generate large volumes of tailings, and surface mines are going to generate overburden because miners have to remove overlying soil and rock to get to the ore. This is not a call for complacency; rather, it suggests that efforts and policies should be aimed at those aspects of environmental degradation over which miners have some control. Examples include efforts to recycle chemical reagents used in benefici-ation, to place or seal waste piles so they are less exposed to the water necessary for acid formation, and to reduce the chances that tailings ponds will leak into surrounding soil and groundwater. Moreover, incre-mental improvements are to be encouraged, both in beneficiation techniques to increase rates of metal recovery and in mining tech-niques to reduce the amount of overburden and waste rock extracted along with metallic ores.

Second, the amount of solid waste generated during mining and beneficiation is not a good indicator of the actual amount of environ-mental damage caused by these activities. The same mineral deposit or mine in different circumstances may generate the same amount of waste but cause substantially different amounts of environmental dam-age because of differences in climate, population density, or one of the other factors noted earlier. The implication for public policy is that rules need to be flexible enough to account for site-specific differences among mines and beneficiation facilities.

Mine Closure and Rehabilitation

A mine eventually reaches the end of its useful life, either because it physically depletes its ore or because economic conditions become unfavorable (costs rise or mineral prices fall). When this happens, *mine closure* and *rehabilitation* (or reclamation) occur. Underground mines typi-cally are sealed or plugged. Surface mines, as well as waste sites for both underground and surface mines, are rehabilitated. Pits and waste piles have their slopes stabilized and may be revegetated. In some cases, acid mine drainage continues even after mining stops, requiring some type of drainage control.

The precise nature of mine closure and rehabilitation varies from place to place because of different public policies and accepted indus-try practices. More fundamentally, closure and rehabilitation activities vary because potential damages from closed mines vary considerably for all the reasons cited earlier, such as type of mine, climate, and prox-imity to population centers.

Key issues for public policy in this area are: the rehabilitation requirements; the mechanism for ensuring that appropriate closure and rehabilitation procedures are followed; and the nature of postmining liability. The rehabilitation requirements often are only vaguely or qualitatively defined: for example, mined land must be returned to a "'usable condition,' to a 'stable condition,' to the 'greatest degree,' or 'equal to the level of highest previous use'" (Intarapravich and Clark, forthcoming). Moreover, economic considerations often seem to get short shrift when rehabilitation standards are being determined; the result can be standards levels for which the costs of rehabilitation exceed the associated benefits. Barnes and Cox (1992), for example, discuss the apparent "goldplating" of rehabilitation requirements in Australia.

The most common mechanism for ensuring that appropriate procedures are followed is a reclamation bond or fund. With a reclamation bond, mining companies put money into an escrow account or in some other way set aside money as a guarantee that they will perform the required reclamation work. The only way for a company to get the money back is to perform the required work. A critical issue for public policy is the size of the bond: high enough to ensure that mine operators actually perform the required work rather than simply forfeit the bonded money, but not so high as to discourage mining.

The nature of postmining liability is perhaps the most controversial issue affecting mine closure and rehabilitation. Subsidence and contaminated mine water are the most common sources of postmining damages for which companies may be required to pay fines or compensation. That companies should be liable for damages caused by nonprudent, negligent, and illegal activities is not controversial. But public policies sometimes define liabilities more broadly. In some cases, a company can be held responsible for damages caused by subsidence and polluted mine water, even if it acted prudently and within existing laws and regulations or was only partially responsible for the damages. A company sometimes can be held responsible for damages retroactively, following changes in laws and regulations. These broader aspects of liability are designed to encourage companies to go beyond simple compliance with existing regulations. But such broad liability has been criticized for being unfair and for discouraging investment in mining activities.

The Lecturers and the Lectures

Against this backdrop of the stages in the life of a mine, the five lectures in this volume assess public policies toward mining and the environment.

Anthony Cox is a senior economist in the Minerals and Energy Group of the Australian Bureau of Agricultural and Resource Economics. He has written extensively on key aspects of the mining industry in Australia. He has been deeply involved in the development and analysis of government policy affecting the mining industry, particularly with respect to environmental issues, taxation, industry assistance, and minerals trade. He undertook his postgraduate training in economics and econometrics at the Australian National University.

In his lecture, Cox focuses on the issue of land access for mineral exploration and mine development. A major source of controversy is the means by which alternative uses of land are valued and compared. (The range of alternative uses includes mining, grazing, timber production, recreation, and preservation, although only rarely would all of these be potential uses for a particular tract of land.) Cox evaluates the appropriateness of using a benefit-cost framework for comparing alternative uses of land, especially when mining and environmental preservation are being considered. Cox uses as an example the recent debate over mineral development at Coronation Hill in the former Kakadu Conservation Zone of the Northern Territory, Australia (the Conservation Zone was incorporated into the larger surrounding Kakadu National Park in 1991).

A benefit-cost framework, as Cox discusses, requires estimating the benefits and costs associated with each alternative use (or combination of uses) and then discounting these values to the present to account for time value. The land use with the highest positive net present value should, in theory, be selected.[4]

---

[4]For many activities, including developing a mine and preserving an environment, we receive benefits and incur costs over a period of time. Benefits and costs in this context include not only monetary values but also nonmonetary values, such as damage to human health from dirty air or water and the pleasure we get from a scenic vista. Discounting is a process for reducing future values (benefits and costs) to make them directly comparable with present values. We tend to value the future less than the present because we are impatient—for example, we would rather receive a dollar today than next year. More specifically, receipts received today are worth more than similar future receipts because they can be invested today and grow in value between now and the future period. Some observers, however, question the appropriateness of discounting, especially when environmental values and long time frames are involved; see Toman (in Darmstadter 1992, 17–18).

Net present value is the difference between total benefits and total costs, appropriately discounted to account for time value.

This sounds simple enough, but in practice it is difficult to implement. Cox discusses the problems associated with valuing goods for which there are no markets (such as clean air or water) and valuing environmental damage that may be irreversible. He notes the problems of dealing with both uncertainty about the future and lack of information on mineral potential and the value of conservation in areas under dispute. More fundamentally, he acknowledges that some people question the appropriateness of the process of discounting in the first place. The Coronation Hill example is noteworthy, among other reasons, for its use of the contingent valuation method to estimate the value people place on preserving environmental quality.[5] The example shows that Australians place significant values on preservation. These values, Cox contends, should not be ignored just because they are difficult to quantify.

Cox concludes that, despite its many problems, benefit-cost analysis is an appropriate—even indispensable—tool for focusing public debate about land use. Those designing public policies need to place greater emphasis on the collection of information about benefits and costs and then on the use of this information in a review process that is open and transparent to all interested parties. He argues that the process by which land-use decisions are made is critically important. Balancing desires for mineral development with those for preservation and conservation will never be easy, but a decision-making process that is open and transparent will promote better decisions.

John Tilton is the William J. Coulter Professor of Mineral Economics at the Colorado School of Mines and a University Fellow of Resources for the Future Over the past twenty years, he has written widely on various economic and policy issues confronting the metal industries, including cyclical instability, material substitution, long-run trends in metal demand, and mining and the environment. He received his Ph.D. degree in economics from Yale University.

Tilton focuses on the later stages of mining and beneficiation and of mine closure and reclamation—specifically the issue of who should pay

---

[5]Contingent valuation methods involve the use of surveys to estimate the value that people place on environmental quality. A survey might, for example, focus on how much people would be willing to pay to preserve a natural environment or to reduce the health risks associated with some type of industrial activity. For an overview of contingent valuation and other methods for valuing environmental damages, see Cummings, Brookshire, and Schulze 1986; Kopp and Smith 1993.

for cleaning up mine wastes—with a perspective from the United States. He notes that what has become known as the polluter-pays principle is the starting point for most discussions in this area, not just in mining but for most polluting activities. Requiring that polluters pay for cleanup encourages efficiency: it requires producers to acknowledge and bear the environmental costs of production, along with the costs they normally bear for capital, labor, raw materials, and other factors of production. It provides incentives for producers to develop and adopt new and cheaper technologies for protecting environmental resources. Requiring that polluters pay is also arguably equitable or fair: those who pollute the environment and benefit from pollution should pay for cleanup.

For ongoing mining, Tilton argues that the polluter-pays principle is quite reasonable as a basis for public policy. But he contends that the principle is inappropriate for dealing with the problems associated with past mining. Many former polluters no longer exist, and others are not financially able to pay for cleanup. Many sites have changed ownership several times over the years, and apportioning blame among the various owners is difficult, if not impossible. Ultimately, Tilton contends, the consumers of goods whose production generated the pollution are responsible for the pollution.

Despite these problems, U.S. policy toward cleaning up hazardous old mining sites relies on the polluter-pays principle. The result, Tilton contends, has been inefficiency in the remediation of sites. Probably between one-fifth and one-third of the total amount spent on remediation has gone toward litigation and other transaction costs associated with determining who is responsible for cleanup (that is, in assigning responsibility for blame). The nature of legal liability in U.S. policy—strict, joint-and-several, and retroactive—discourages remining of old mining and milling sites, often the least expensive method of cleanup. (The three forms of legal liability are defined and their implications are discussed in Tilton's lecture.)

The remaining three lectures do not concentrate on specific stages of production, but instead provide perspectives that cut across all three stages. Gustavo Lagos is a professor of mining engineering at the Pontificia Universidad Catolica de Chile and former executive director of the Center for Copper and Mining Studies, a research organization in Santiago, Chile. He is a member of the Board of Directors of the Center for Environmental Research and Planning and has been a consultant for the government Commission on Mining and the Environment. He received his Ph.D. degree from Leeds University, United Kingdom.

Lagos concentrates on formulating national environmental policies in the developing world, using Chile as an example. In the United States and other industrialized countries, the recent discussions about public policy toward mining and the environment have taken place in the context of existing national policies and regulatory mechanisms dating back in many instances to the 1960s and 1970s. Many mineral-rich developing countries have had little in the way of national environmental policies. In other such developing countries, environmental laws are on the books, but are not effective because of conflicting objectives in different policies, bureaucratic confusion, and weak enforcement; Chile is an example.

Lagos notes that, until very recently, Chile did not require environmental impact studies prior to issuing permits for large projects, had no requirements for land reclamation following mining, and had no standards regulating liquid and solid effluent or soil quality. But this is changing. In less than a decade, public awareness about the environment has grown significantly, and public policy is racing to catch up. A national framework for environmental policy, proposed in 1992, rests on three principles: legal liability for environmental damages, absent previously; environmental impact studies for large new projects affecting the environment; and polluters paying for cleanup. With these principles as a framework, Lagos expects that national legislation will be developed by the end of the 1990s governing, for example, water quality, soil quality, abandoned mines, and tailings dams.

Despite the infancy of environmental policy in Chile, environmental practice is in some cases ahead of national requirements. This is especially true for new projects. In the absence of environmental policies or standards, foreign investors typically design and construct facilities to meet the standards of their home countries (such as the United States or Japan), both in anticipation of the development of stricter standards in Chile and to avoid shareholder accusations of exporting pollution to the developing world. If true not only in Chile but elsewhere in the developing world, the proposition that industrial activity in general and mining in particular will migrate to countries with less strict environmental regulation is weaker than often argued.

The environmental practices agreed upon in ad hoc discussions between foreign investors and local Chileans are likely to significantly shape later legislation and regulations. Thus environmental policies of countries like Australia, Japan, and the United States are likely to have an important influence on future policies in developing countries through their "demonstration" effect. But Lagos argues that it would not be wise for Chile to blindly mimic the environmental policies of the United States and other countries with several decades of experience with environmental policy. Chilean policies should be developed gradually, learning from the experiences of other countries and adapting existing policies elsewhere to the Chilean situation.

David Humphreys is deputy chief economist at the RTZ Corporation, an international mining company based in London. He has written extensively on the metals and minerals industry. Prior to joining RTZ, he worked in U.K. government service for nine years. He is vice president of the Brussels-based industry federation, Euromines, and associate editor of the journal *Resources Policy*. He received his Ph.D. degree in economics from the University of Wales.

Humphreys assesses Europe and its attempt to elaborate supranational policies toward the environment. Supranational organizations and multinational agreements have become increasingly important in environmental policy; consider, for example, the 1987 Montreal protocol on chlorofluorocarbons; the biodiversity convention and other agreements of the 1992 U.N. Conference on Environment and Development (known as the Earth Summit); efforts by multilateral development banks (such as the World Bank) to promote improved environmental practices; and inclusion of (very general) environmental provisions in the North American Free Trade Agreement.

Humphreys observes that, to date, European Community (EC) legislation on the environment has had little effect on the mineral industries; national policies have been more important. He nevertheless concludes that EC influence on environmental matters is bound to increase. For mining, greater EC influence is likely to mean tougher requirements for environmental impact assessments as part of preproduction planning and permitting. Mine wastes, from both ongoing operations and from the past, also are likely to be targets of EC policy.

Humphreys notes the tension between supranational and national interests—EC desires for common environmental standards and regulations versus national desires to tailor rules to fit individual circumstances. The European solution to this tension is the principle of subsidiarity, by which action is taken at the EC level only if it cannot be achieved by the member nations themselves. The principle seems simple enough but has been prone to controversy in interpretation. The principle and its interpretation lie at the heart of the ongoing debate over European union. The approach emerging, in particular the growing acceptance that the achievement of environmental objectives at a supranational level requires policies to have some flexibility to accommodate differing national development aspirations and constraints, potentially has implications for other attempts at supranational policymaking.

Humphreys observes a gradual change in the tone and philosophy underlying the development of environmental policy in Europe. Most early environmental rules relied heavily on direct regulation, especially technology-based standards and performance standards that defined maximum allowable emissions. Governments developed these rules in a largely piecemeal fashion, in response to specific incidents or problems. The process of policy development was confrontational—government versus business. More recently, Humphreys argues, the process has become less reactive and more proactive, what Humphreys calls an integrated approach to policy development. Although direct regulation still is alive and well, more attention is being paid to the potential offered by economic instruments such as emission taxes. Emerging from the political reality of trying to forge agreement amongst twelve sovereign states is a recognition that government, industry, and society as a whole benefit from an approach to environmental matters that is less confrontational and legalistic and more cooperative in tone.

Alyson Warhurst directs the international Mining and Environment Research Network, which brings together interdisciplinary research teams from around the world to study and disseminate policy analysis about environmental regulation, technical change, and international competitiveness. She also is a senior fellow at the Science Policy Research Unit of the University of Sussex in the United Kingdom. She received her Ph.D. degree in industrial policy studies from the University of Sussex. Warhurst has worked extensively in South America, and her research career has focused on technical change in the extractive industries, including offshore oil and mining, with special emphasis on developing countries.

In the final lecture, Warhurst focuses on the roles of technological innovation and human resources in environmental management, especially in developing countries. Much existing environmental policy emphasizes the modification of existing facilities to reduce emissions (add-on solutions to end-of-pipe problems); companies direct their efforts toward complying with specific regulations and meeting specific standards for emissions or pollution-control equipment. There is little incentive to innovate and develop new and less costly means of achieving a desired level of environmental quality. Instead, the incentive is to comply with existing rules and get on with business.

In the long run, however, new and better methods of mining and mineral processing are the key to reducing both the costs of mineral production and the environmental consequences of mining and mineral processing. Technological and organizational innovation, in other words, offers a potential escape from the oft-presumed trade-off between economic growth and environmental quality. Furthermore, better equipment and techniques will require more highly trained workers.

Warhurst argues that what most people consider to be "environmental policy"—including, for example, permitting and other preproduction approvals, and performance and technology-based standards—is only a partial solution to the growing concern over the environmental consequences of mineral production. She calls for governments to create a business environment in which technological change and human-resource development both can thrive.

Although Warhurst does not focus on a specific country or set of countries, her essay contains important lessons for many developing countries, where much of the environmental degradation from mining and mineral processing can be attributed to obsolete technology and out-of-date work practices. Important issues for these countries, therefore, are gaining access to existing state-of-the-art technologies and providing the education and training necessary to bring managers, engineers, and other workers up to international standards. For mining companies, Warhurst argues, there are competitive advantages to be gained and profits to be made by being innovative in environmental management.

Final Thoughts

This introduction began by quoting Georgius Agricola, writing in the 1500s about the environmental consequences of mining. Were Agricola to reappear today, he would not be surprised that mining continues to disturb the environment. He likely would note that a certain amount of environmental degradation is inevitable if mining—and in fact all other economic activities—is to occur. But in other respects, Agricola likely would be surprised: at the extent to which society no longer accepts the denuded landscapes, fouled streams, and dirty air that in his time were considered the price that had to be paid for mineral production, and at the extent to which mining companies have responded to society's demands for a cleaner environment and made the environment an important consideration in investment and operating decisions.

We hope that Agricola also would write that this collection of lectures in some small way has contributed to the ongoing international debate about public policy toward mining and the environment.

References

Agricola, Georgius. 1950. *De Re Metallica*. 1556. Translated by Herbert Clark Hoover and Lou Henry Hoover. 1912. Reprint, New York: Dover Publications.

Australian Mining Industry Council. 1991. *Mine Rehabilitation Handbook*. Dickson, Australia: Australian Mining Industry Council.

Barnes, P., and A. Cox. 1992. Mine Rehabilitation: An Economic Perspective on a Technical Activity. In *Rehabilitate Victoria: Advances in Mine Environmental Planning and Rehabilitation*. Proceedings of the Australian Institute of Mining and Metallurgy Conference. Publication Series No. 11/92. 149-158.

Cummings, Ronald G., David S. Brookshire, and William D. Schulze, eds. 1986. *Valuing Environmental Goods: An Assessment of the Contingent Value Method*. Totowa, New Jersey: Rowman and Littlefield.

Darmstadter, Joel, ed. 1992. *Global Development and the Environment: Perspectives on Sustainability*. Washington, D.C.: Resources for the Future.

EPA (U.S. Environmental Protection Agency). Office of Solid Waste and Emergency Response. 1985. *Report to Congress: Wastes from the Extraction and Beneficiation of Metallic Ores, Phosphate, Asbestos, Overburden from Uranium Mining, and Oil Shale*. EPA/530-SW-85-033. Washington, D.C.: U.S. Government Printing Office.

The Fifth Discipline. 1993. *Mining Magazine*, December. 290.

Hartman, Howard L., ed. 1992. SME *Mining Engineering Handbook*. 2d ed. Littleton, Colorado: Society for Mining, Metallurgy, and Exploration.

Hutchison, Ian P.G., and Richard D. Ellison, eds. 1992. *Mine Waste Management*. Boca Raton, Florida: Lewis Publishers.

Intarapravich, Duangjai, and Allen L. Clark. Forthcoming. Performance Guarantee Schemes in the Minerals Industry: The Case of Thailand. *Resources Policy.*

Johnson, Wilton, and James Paone. 1982. *Land Utilization and Reclamation in the Mining Industry, 1930–1980.* Bureau of Mines Information Circular 8862. Washington, D.C.: U.S. Department of the Interior, Bureau of Mines.

Kopp, Raymond J., and V. Kerry Smith, eds. 1993. *Valuing Natural Assets: The Economics of Natural Resource Damage Assessment.* Washington, D.C.: Resources for the Future.

Lootens, D.J., W.M. Greenslade, and J.M. Barker, eds. 1991. *Environmental Management for the 1990s.* Littleton, Colorado: Society for Mining, Metallurgy, and Exploration.

Office of Technology Assessment. 1992. *Managing Industrial Solid Wastes from Manufacturing, Mining, Oil and Gas Production, and Utility Coal Combustion: Background Paper.* OTA-BP-O-82. Washington, D.C.: U.S. Government Printing Office.

Sengupta, Mirtunjoy. 1993. *Environmental Impacts of Mining: Monitoring, Restoration, and Control.* Boca Raton, Florida: Lewis Publishers.

United Nations Environment Programme. 1991a. *Environmental Aspects of Selected Nonferrous Metals (Cu, Ni, Pb, Zn, Au) Ore Mining: A Technical Guide.* UNEP/IEPAC technical report series no. 5. Paris: United Nations Environment Programme.

———. 1991b. *Environmental Data Report.* 3d ed. Oxford: Basil Blackwell.

Van Syl, Dirk, Marshall Koval, and Ta M. Li, eds. 1992. *Risk Assessment/Management Issues in the Environmental Planning of Mines.* Littleton, Colorado: Society for Mining, Metallurgy, and Exploration.

World Commission on Environment and Development. 1987. *Our Common Future.* Oxford: Oxford University Press.

Yegulalp, T.M., and K. Kim, eds. 1990. *Environmental Issues and Waste Management in Energy and Minerals Production.* Columbus, Ohio: Batelle Press.

Land Access for Mineral Development in Australia

ANTHONY COX

The issue of access to land for exploration and mining remains one of the major policy concerns facing the mining industry, conservation groups, and Commonwealth and State governments in Australia. Exploration for or development of mineral resources is only one of several possible uses for a given area of land. Other uses include conservation, Aboriginal heritage, recreation, farming, forestry, and urban settlement. In some areas, these uses will be mutually exclusive, while in other areas, some combination of uses will be possible. Moreover, it may be possible for alternative uses to be made of the same tract of land over time.

In general, most exploration and mining projects in Australia are approved with little contention. However, conflicts over access to land for mining and exploration have become more frequent and more public over the past decade. Despite some positive moves made recently at the Commonwealth and State government levels, existing institutional arrangements are not providing an effective and efficient means of resolving conflicts between mining and conservation uses of land. Current decision-making processes are often perceived as being ad hoc and inconsistent by both the mining industry and conservation groups.

The reasons for such conflicts are complex. Most of the major conflicts between mining and conservation groups tend to involve areas of land that have significant competing values and attract wide public interest. The conflicts can also involve a range of environmental impacts, ethical and valuation concerns, property rights issues, risks and uncertainties, and pressures for policy change, all of which complicate public policy processes. However, there is scope to make existing institutional arrangements in Australia more effective and efficient in both avoiding and resolving conflicts between mining and conservation uses of land.

In this paper, the policy problem of access to land in Australia is placed in an economic perspective. The next section describes the policy problem and attempts to place it in a broader policy context. Attention is then paid to the issues underlying the policy problem, particularly the issues relating to defining significant environmental impacts, defining property rights, the lack of adequate and appropriate information for decision making, ethical and valuation concerns, and irreversibility. A decision-making framework based on benefit-cost analysis is then discussed, using the recent debate over mining development at Coronation Hill in the Northern Territory to illustrate a particular attempt to resolve land access conflicts.

The Policy Problem in Perspective

Governments face a fundamental problem when managing natural resources: how to ensure that decisions about the use, or combination of uses, to which particular areas of land are dedicated will maximise the net welfare of society over time. From a social perspective, contributions to welfare stem from the full range of values society derives from resources and their uses. Use of natural resources in this sense includes both commercial and conservation aspects of production and consumption. As well as providing the base for Australia's major export industries and for important domestically oriented sectors of the Australian economy, natural resources provide a wide range of environmental benefits. These include access to clean air and water, recreational benefits, preservation of genetic diversity, and existence benefits.

At the broad conceptual level, the policy problem seems relatively straightforward. The issue of importance to governments is whether decisions about land use allow for competing values to be genuinely assessed. Only if the relative values of different land use options are carefully weighed will the community's resources be allocated to the uses that maximise national welfare. Decision-making mechanisms therefore need to allow the relative costs and benefits of different land use regimes to be effectively compared within the constraints of attainable information and available valuation techniques.

The Australian debate over land access has generally been cast in terms of how much land is available for exploration and mining, both now and in the future. The mining industry, for example, claims that around 26 percent of the Australian land mass is effectively sterilised from exploration and mining. This figure consists of national parks and conservation areas where exploration and mining are virtually prohibited (5.6 percent), Aboriginal areas where access is restricted and subject to special conditions (15.3 percent), rural land subject to owner veto on exploration (2.7 percent), and other areas subject to varying degrees of restriction, such as urban areas and forestry reserves (2.3 percent) (Australian Mining Industry Council 1990).

Problems flowing from restrictions on land access are deemed by the industry to be significant. According to the Association of Mining and Exploration Companies, 'without guaranteed access to land for mineral exploration purposes, on reasonable and practical terms, Australia's mining industry would begin to decline, to a point some years hence when as a viable industry it would cease to exist as an economic force' (Industry Commission 1991, 2). Zimmerman (1992) claims that more and more Australian enterprises are moving much of their exploration and development capital offshore, as frustrations, impediments, and delays increase.

The environmental movement views these figures with some scepticism. The Australian Conservation Foundation (ACF), for example, interprets the figures very differently and claims that the mining industry has exaggerated the restrictions imposed on it. By ACF's analysis, less than 7 percent of Australia is closed to or difficult to access for mining (Krockenberger 1992). ACF's figure consists of the amount of land contained in national parks and urban areas. ACF claims that the rest of Australia is available for exploration and mining, albeit with some negotiations required for gaining access. ACF goes on to note that 'it is hard to believe that Australia's mineral future is dependant [sic] on that 7 percent. If it is, the outlook for the industry is bleak' (Krockenberger 1992).

From a public policy perspective, however, this debate about the proportion of land available for mining or reserved for national parks is misdirected and unhelpful to the decision-making process. Whether the amount of land currently devoted to national parks or to mining is socially optimal is a question to which there is unlikely to be an easy answer. The social optimum is almost certain to change over time in accordance with changes in social preferences, income, and economic

conditions. Clearly, access to land for exploration and mining is necessary if the industry is to expand the mineral resource base and replen-
ish the reserves currently being depleted. What is equally clear is that
conservation of natural areas is becoming increasingly important to
large sections of the Australian community. The fact is that often the
same small percentage of land is highly valued by both miners and
conservationists. Examples abound of regions that are highly prospective for minerals but are also of high ecological significance: the Kakadu
region in the Northern Territory, the tropical rain forests of north
Queensland, and the southwest wilderness area of Tasmania.

How significant, then, is the land access problem in Australia? On
one hand, there is little evidence that many mining projects have been
stopped on purely environmental grounds. On the other hand, the
project approval process, which incorporates environmental impact
assessment, has undoubtedly resulted in the way in which some projects are developed being adjusted to meet environmental constraints,
with attendant delays and increases in costs to the industry. In general,
though, the implications of making poorly informed decisions on land
uses are significant. Most obviously, there will be a loss in national welfare if particular land management regimes are implemented that arbi-
trarily or accidentally exclude activities which would have increased
welfare. One consequence of particular concern is the perception that
sovereign risk attached to exploration and mining development in
Australia may increase as a result of perceived instability and uncertainty in government approval processes. Perceptions of increased sovereign risk in relation to environmental policies could adversely affect
the competitiveness of Australia as an investment location compared
with other countries with similar mineral potential.

It is doubtful that the overall risk premium for exploration expenditure in Australia is greater than that, for example, in North America, and
it is also probably less than that in most countries in Asia and South
America. In North America, for example, similar concerns have been
raised about access to land, and there have been calls for changes to
government policies (Eggert 1989). There are also factors other than
sovereign risk that play a significant role in influencing decisions about
exploration and development investment. Johnson (1990) concluded
that geologic potential and political stability rank higher than government mineral policy (sovereign risk) when companies are deciding on
the location of exploration and development investment. Infrastructure
availability, past experience in a country, present and expected min-

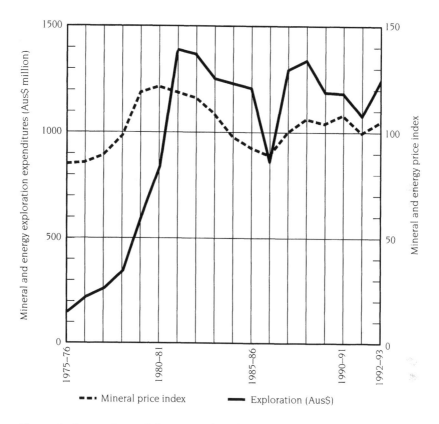

Figure 1. Comparison of the mineral and energy price index and expenditures on mineral and energy exploration in Australia, 1975–1993.
Source: ABARE 1993b, Table 28

eral prices, mining industry profitability, and corporate goals are among other influencing factors. Mineral prices in particular appear to have had a strong influence on exploration expenditure in Australia over the past twenty years (fig. 1). The removal of the company taxation exemption for gold production in 1991 also helps to explain the level of and shifts in the exploration targets of Australian companies (fig. 2). Gold has been, and continues to be, by far the dominant nonfuel mineral exploration target. While it is apparent that Australian mining companies are expanding investment in exploration and development overseas, in significant part this reflects the increasingly multinational

25

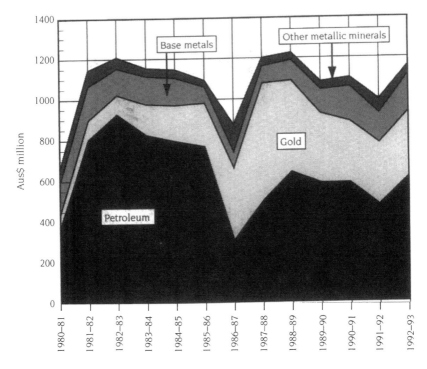

Figure 2. Cumulative expenditures for exploration of petroleum, gold, base metals, and other metallic minerals in Australia, 1980–1993.

Source: ABARE 1993b, Table 28

nature of the Australian mining industry and is not necessarily linked to the issue of perceived differentials in sovereign risk.

There is no doubt, however, that mining industry perceptions of government policy and sovereign risk do have an influence on the level and location of exploration investment. Increased risk arising from slow and poor decision-making processes on the part of government is an avoidable cost burden on both the mining industry and the community as a whole. The aim of efficient public policy should be to provide certainty about the process by which decisions like those concerning resource use are made. This is not to say that the actual outcome of the decision-making process should be a foregone conclusion, but that the institutional mechanisms and the processes and criteria by which decisions are made should be clear, objective, and efficient.

Coronation Hill

The history of the proposed gold, platinum, and palladium mine at Coronation Hill in the Kakadu Conservation Zone, located in the Northern Territory, provides an interesting example of recent Australian experience with land access issues. In 1991, the Resource Assessment Commission (RAC) completed an inquiry into the proposed mine and the policy issues surrounding the alternative uses of the Kakadu Conservation Zone (Resource Assessment Commission 1991a). The RAC was established by the Commonwealth government in 1989 to deal with highly contentious environmental and resource allocation issues. The RAC was established as a statutory body that would independently and objectively hear and assess all of the claims to the use of a resource, with the aim of integrating the economic and conservation aspects of development. The inquiry process relied heavily on public hearings and submissions from interested parties. At the time, the RAC was heralded as a significant institutional innovation in environmental dispute resolution and an experiment in transparent policy analysis (Galligan and Lynch 1992, 3).[1]

The Kakadu Conservation Zone was an area of land surrounded by, but not a part of, Stage 3 of the Kakadu National Park. Stages 1 and 2 of the Park had been proclaimed in 1979 and 1984 respectively. The Conservation Zone comprised an area of 47.5 square kilometres and was set aside when Stage 3 was declared in 1987, with the knowledge that it contained a proposed mine, to enable the mineral potential of the area to be assessed prior to a decision being made on whether to incorporate the area into the larger (19,760 square kilometres) Kakadu National Park. The site of the proposed Coronation Hill mine was in the Conservation Zone adjacent to the South Alligator River, some 100 kilometres upstream from the World Heritage wetlands of Kakadu National Park.

Pastoral activity and mining had proceeded intermittently for half a century on large sections of the park surrounding the Conservation Zone, and there was little evidence that these economic activities had led to any damage to the downstream wetlands. Uranium had been discovered at Coronation Hill in the 1950s, and known deposits were extensively mined by small-scale operations for a few years. In 1984, the Coronation Hill Joint Venture (comprising BHP Gold Mines, Pioneer

---

[1]The RAC was disbanded in December 1993.

Minerals, and North Broken Hill Holdings) began drilling for gold and also found platinum and palladium mineralisation. The final environmental impact statement was submitted to the Commonwealth government in 1989, seeking approval for the mine to proceed. Although the statement satisfied most of the environmental requirements, the government deferred a decision on the mine because of concerns regarding Aboriginal claims in the area and the cumulative environmental impact that could result if further mines were allowed in the Conservation Zone. The government finally referred the matter to the RAC in 1990.

The controversy over development of the mine involved mining, conservation, and Aboriginal interests. Each group attached a great deal of significance to the process and the outcome of the RAC inquiry. The inquiry process took twelve months to complete, and the final report was based on written submissions, public hearings, meetings with interested parties, consultancy reports, and examination of relevant published and unpublished information. The decisions faced by the Commonwealth government were widely expected to establish precedents for the way in which mining and conservation claims to natural resources would be approached in the future. All groups involved in the dispute viewed the process as a test for the government's environmental, developmental, or cultural credentials. From the government's perspective, it was hoped that the RAC would be capable of developing a policy framework for handling difficult development questions and of clarifying the options for resource allocation.

In its final report, the RAC found that the environmental effects of mining the Coronation Hill site would not threaten the environmental balance of the fragile wetlands of Kakadu. Specifically, the inquiry found that, with appropriate environmental safeguards, the impact of mining would not cause significant harm to the water flow levels in the South Alligator River. The report signalled, however, that mining would threaten the integrity of the park's World Heritage listing and would irretrievably damage Aboriginals' spiritual values, as the area sought for mining is claimed by the Jawoyn tribe to be a sacred site. After the RAC presented its report, the Commonwealth government decided to incorporate the Conservation Zone into Kakadu National Park. This decision was made not on the basis of environmental concerns, but on the basis of difficult questions arising from the claims of Aborigines to areas of spiritual significance within the Conservation Zone. Nevertheless, the process of the inquiry shone a spotlight on the role of economics in assessing mining and conservation claims to land. This experience will be discussed in further detail below.

Land Access Issues for Policymakers

There are a number of issues that are fundamental to the land access debate and that must be understood by policymakers before an integrated policy approach to the problem can be undertaken. These include issues associated with defining significant environmental impacts, defining property rights, the process by which conservation areas are declared, the lack of appropriate information, ethical and valuation concerns, and irreversibility. The major challenge facing governments in resolving land access conflicts lies in recognising and trying to understand the significance of these major issues. From a process that takes them effectively into account, better, more informed resource management decisions can be made. While each of these issues is linked in some measure to the others, they will be considered in turn to provide an exposition of the difficulties facing policymakers.

Environmental Impacts

A first step in analysing the nature of the conflict between mining and conservation uses of land is identifying the types of environmental impacts associated with mining. Here, an important distinction needs to be made between the exploration for minerals and the development of mines. While all mining activities affect the surrounding environment in some way, the magnitude and type of these impacts can vary greatly among the stages of mining. This variation has important implications for the development of public policy dealing with land access.

In general, exploration causes the least disturbance to the environment in the exploration-mining-processing sequence. Some forms of exploration have no environmental impacts (for example, remote sensing by satellite). Others, such as airborne surveys, involve only minor temporary effects, such as noise pollution. Still other exploration activities, such as drilling and exploratory stripping and trenching, may result in slightly greater environmental disturbances.

Using different and modern forms of exploration techniques, many of the potentially adverse environmental impacts of exploration can be prevented, albeit at a cost. For example, the increased use of heli-copters reduces the need for access roads to be constructed. While exploration may detract somewhat from other uses of a piece of land (for example, by damaging some conservation values), it is only in a few cases that exploration on land will completely rule out other uses such as conservation. However, the issue is not whether these impacts of exploration can be minimised in every case, since they can always be entirely avoided by banning exploration. Rather, the issue is to determine what level of disturbance and damage to the environment is socially acceptable, given the potential gains associated with mineral exploration and the environmental sensitivities of the land proposed to be the subject of exploration activity.

In contrast to exploration, mining itself generally involves greater disturbance to the land surface and is therefore generally seen as pos-ing greater difficulties in terms of compatibility with other uses. The development stage of mining involves sinking shafts, excavating pits, and constructing buildings and transport facilities. Dust, noise, and other environmental as well as socioeconomic impacts are generated by the relatively large and sudden influx of people involved in a min-ing operation. Surface disturbance and solid waste generation are usu-ally unavoidable at this stage. Processing minerals on-site usually involves crushing and grinding operations, which may generate air-borne particulates and noise. Waste disposal problems are associated with tailings, which may be distributed by wind and water erosion over the areas surrounding a mine.

As with exploration, many environmental effects of mining activi-ties can be mitigated, but only at a cost. Dust suppression techniques and collection systems can be used to reduce particulate emissions. Water pollution, which probably represents the major potential threat to the environment from mining, can be controlled through a variety of in-process and end-of-process techniques. In most cases, rehabilitation of mine sites at the completion of operations can restore many of the features of the landscape that existed before mining began, assist in the re-establishment of ecosystems, and reduce the potential for pollu-tion to flow from the abandoned mine site in the future. Indeed, Australian mining companies are at the forefront of world technology for mine rehabilitation; for example, Alcoa was placed on the United Nations Environment Programme's Global 500 Honour Roll for achieve-ments in environmental management and rehabilitation at its bauxite mines in southwest Australia.

From a policy perspective, a distinction could therefore be made between access to land for exploration and access for mining. Exploration has much less environmental impact than mining, moving relatively quickly from large tracts of land to narrow areas of higher prospectivity. Exploration can primarily be seen as an information-gathering activity that provides input to the decision-making process about the relative values of land use.

Property Rights Issues

Often the debate between mining and conservation groups has been cast in terms of whether exploration and mining should dominate conservation or whether this dominance should be reversed.[2] It can be argued that, traditionally, mining, together with agriculture, has been seen as a dominant land use in Australia (Bolton 1981). At least until the late 1960s, mining interests in Australia were generally unhindered in gaining access to land for exploration and development. Implicit in this history is the presumption that mining is a land use that is in society's interest to encourage, because the resulting benefits outweigh the costs.

In more recent times, however, this presumption has been increasingly questioned, both because of higher values being attached to competing land uses and because of concerns that all of the costs of mining have not necessarily been taken into account. This re-evalution is critically affected by the way in which rights over valuable mineral and land assets are specified. The nature of ownership rights to resources can affect efficiency of resource use as well as the distribution of wealth and income. If ownership rights are well defined and transferable, then profit-maximising resource use can be efficient, provided the prices of all costs and benefits of using resources reflect social values. In Australia, as elsewhere, policy decisions are influenced by the ownership and allocation of mineral rights and by the difficulties associated with adequately defining property rights to many goods and services provided by the environment.

---

[2]The argument that mining should take precedence over other land uses is often based on the notion that the location of mineral deposits has been fixed by geological events of the past. However, a similar argument could also be mounted that areas of high conservation value have been determined by natural processes of the past.

Ownership of Land and Minerals. In Australia, ownership of mineral resources generally lies with the Crown (in practice, State, Territory, and Commonwealth governments), regardless of who owns the land on the surface. This reflects a belief that mineral deposits are a fortuitous gift of nature and that any net benefits flowing from their exploitation should accrue to the community as a whole, rather than to whoever happens to own the surface rights.[3]

An important feature of the Australian system of Crown ownership is that the Commonwealth government is not the principal holder of mineral rights. Since mining is not explicitly mentioned in the Australian Constitution, ownership of minerals found onshore and offshore within the three-nautical-mile territorial limit defaults to the relevant State or Territory government. Minerals found beyond the three-mile limit or in external territories are the property of the Commonwealth government. This division is more a product of historical accident than based on any underlying principles of equity.[4]

Other aspects of the Australian Constitution also serve to blur the division of mineral rights between the levels of government. Commonwealth power over matters such as international trade, taxation, defence, people of any race, and external affairs (including international treaties such as the World Heritage Convention) can be exercised in some cases to severely restrict the rights of State or Territory governments to use as they see fit the resources over which they have claim. For example, the Commonwealth government's imposition for environmental reasons of export controls over mineral sands products effect-

---

[3]There are isolated exceptions to public ownership of mineral resources. In New South Wales, Aboriginal Land Councils have been granted title over minerals on their land (with the exceptions of gold, silver, coal, and petroleum). In addition, a few mining leases granted before 1899 that are still operational, such as the Hampton Plains Estate in the Kalgoorlie region of Western Australia, provide for private ownership of minerals except for the royal minerals of gold and silver. This reflects the fact that in the early years of settlement, British common law applied, and all minerals (except for gold and silver) belonged to the landowner. Governments in Australia have gradually reversed the common law position by progressively adopting a practice of reserving minerals from land grants, and this now applies to all minerals.

[4]There are exceptions to this general division of property rights between the State and Commonwealth governments. The most important of these occurs in the Northern Territory, where the Commonwealth government retained property rights over uranium and other substances prescribed in the Atomic Energy Act 1953 following the granting of self-government to the Northern Territory.

ively constrained the rights of the Queensland Government to use the mineral sands resources on Fraser Island. In recent times, the Commonwealth government has increased its involvement in resource development through such indirect means. In particular, increased use has been made of international treaties to achieve environmental objectives in Australia.

Allocation of Mineral Rights. Development of mineral resources takes place under the *regalian* system, where the state owns all mineral resources and leases rights to explore for and extract these resources to private firms under set conditions.[5] Conditions covering exploration can include, for example, annual rental payments, minimum annual expenditure or work requirements, a phased relinquishment of land if nothing is found within specified periods, and regular reporting of geological information gathered. In most States and in the Northern Territory, exploration licences are granted on the basis of 'first come, first served.' As the name implies, under this system exploration rights over a given area are allocated to the first applicant. Simultaneous applications may be resolved by ballot or, more usually, by assessing work programs.

Conversion of a right to explore into a right to mine generally requires the satisfaction of a number of additional conditions. Importantly, this conversion is not automatic and may be subject to compliance with a range of further conditions (for example, minimising adverse environmental consequences), some of which may not be known in advance. The holder of a right to explore is normally given the first option over mining rights. In return for the transfer of property rights to the private sector, it is generally accepted that governments, on behalf of the communities they represent, should be compensated. Since governments in Australia typically allocate exploration and mining rights on a nonprice basis, full payment is not received at the time of their allocation. Royalties, resource rent taxes, and other payments (such as excess government charges for rail services) are instruments that are used to ensure that the community receives a return for the exploitation of publicly owned resources.

Clearly, it is of no benefit to the mining industry to have the rights to exploit mineral deposits if they cannot get to them—that is, without

---

[5]There is some public development of resources, such as brown coal in Victoria, and some minerals are owned by private landholders in Tasmania.

some right of access to the land above a mineral deposit. If the owner of the resource is also the owner of the land (for example, the relevant State, Territory, or Commonwealth government), negotiations for access to land and minerals can be handled simultaneously. Negotiations become more complicated where surface and subsurface rights are held separately. While the legal framework differs marginally among the various jurisdictions, some generalisations can be made. Where a potential developer desires access to unoccupied Crown land, the main considerations in determining whether an exploration or mining lease will be granted relate to the existence of conflicting public uses of the land, such as for forestry or national parks. In the case of occupied Crown land or private land, a mining lease may be granted, provided agreement has been reached between the miner and the current occupier for compensation for any damage caused by the mining operation. In the event of a disagreement on the amount of compensation to be paid, either party can have the issue settled by a court ruling.[6]

It should be remembered, however, that enhancement of mining is not—as often assumed by the mining industry—necessarily efficient from society's point of view. It is possible, for example, that the system of rights established through Crown ownership has led to situations where mining has taken place on some land at a given time and in a given manner even though, after taking into account all the social and environmental costs and benefits, society would have been better off had that land remained in an alternative use (such as agriculture or conservation) for at least some period. It is also possible, of course, that some mineral deposits that should have been mined have not been, due to inefficient land management decisions. Separation of mineral rights from land title complicates the land access issue because there is then a need for some external mechanism to take into account the costs and benefits of decisions made by the owner of one asset (for example, minerals) on the owner of the other (land). If this is not done at all, or done inadequately, inefficient resource use decisions can result.

---

[6]There are some exceptions to the general rule that private landowners have no right of veto over mineral development on their property. Of most relevance are exemptions accorded to owners of cultivated or otherwise improved land in some States and special consent procedures in relation to Aboriginal land in the Northern Territory, which were spelt out in the Aboriginal Land Rights (Northern Territory) Act 1976.

Environmental Concerns and Property Rights. The voluntary transfer of private property rights in a free market may not necessarily result in a socially optimal use of resources. A market approach to allocating resources tends to run into problems when it is technologically difficult or costly to define or enforce property rights. Under these circumstances, resources may be prevented from flowing to their most highly valued use because of the lack of a basis for trade or because the costs of negotiating mutually beneficial outcomes are prohibitive.

A significant example of this problem lies in the difficulty of adequately specifying property rights for many of the goods and services provided by the environment. For a market to behave efficiently and allow for mutually advantageous trade where the benefits and costs of ownership or access are internalised to participants in the trade requires the existence of a pure private good, that is, a good with the characteristics of exclusiveness and rivalry in consumption. A good or service is said to be exclusive if it is possible for the owner to prevent others from consuming the good or service. Rivalry occurs where consumption of the good or service by one person diminishes the amount available to others. A simple case in point is existence value, where it is difficult to exclude individuals from obtaining benefit from the knowledge that a species or ecosystem exists. Similarly, the enjoyment that an individual gains from this knowledge does not reduce the enjoyment that any other individual may receive from that knowledge.

As a result, it is not straightforward to devise solutions to conflicts between mining and conservation interests based on assignment of property rights and subsequent trade between economic agents.[7] In principle, it may not be difficult to establish a property right to an area of land, but the allocation of that right through, for example, an auctioning process would require a bid from all parties who value the resource—that is, both from the beneficiaries of mineral products and from the beneficiaries of conservation. However, it is clear that the social preferences and values regarding conservation will not be adequately represented in any actual bidding process. The costs of organising a bid by conservation interests and collecting actual payment

---

[7]In cases where natural resources have significant conservation benefits but for which recreational benefits are also important, markets are likely to provide for some conservation. For example, private interests may bid for national parks that have a high level of tourist appeal. The recreational benefits that are derived from this are nonrival but exclusive in nature and are therefore amenable to a market solution.

from conservation beneficiaries would be prohibitive because the beneficiaries of conservation are widely dispersed and individuals would have a strong incentive to free-ride on the contributions of others. There are, however, cases where private interests work in favour of conservation without the need for government intervention. Grove (1988), for example, describes the involvement of the Nature Conservancy, a U.S.-based nonprofit agency, in the conservation of wildlife and plant populations.

In the case of land with significant conservation value, therefore, the market may fail to value the assets packaged in that land appropriately, since the value to any one person is less than the value to society as a whole. The market, left to itself, may not lead to the best use being made of the community's scarce resources. This highlights the need to examine the scope for government intervention to determine the appropriate allocation of resources and the likelihood that the net benefits from intervention will be positive.

Declaration of Conservation Areas

The mining versus conservation debate is perhaps at its most divisive when the issue of exploration and mining in national parks is raised. On one hand, some conservation groups argue that virtually all economic activity should be banned in national parks and similar areas. Others concede that the debate should be about the net present values generated by alternative land uses irrespective of their location on the continuum between national parks and cities. There would be overall benefits to Australia from an approach that recognised these trade-offs when declaring national parks and distinguished between different types of activity (for example, exploration as opposed to mining). It is not clear that the current system used to assess the biological, mineral, or other natural resources of proposed conservation areas in Australia incorporates these considerations.

At the Commonwealth level, the National Parks and Wildlife Conservation Act 1975 applies in areas of its jurisdiction. There is separate nature conservation legislation at the State and Territory level. The procedures are similar at both levels of government. In summary, the Australian National Parks and Wildlife Service is required to give public notice of any proposal to declare a park or reserve and to allow sixty days for public comments (to which due consideration must be given). A report on the proposal and the public representations is then prepared and forwarded to the parliament for consideration. The mining industry and other interest groups can make their views known through the public consultation process. The process allows for the alteration of boundaries and even the abandonment of the proposed declaration if counter-arguments are sufficiently strong. Once a national park is declared, exploration and mining are prohibited in the area, or only allowed subject to approval by both houses of parliament in the relevant legislature.

Even where exploration or mining in proposed conservation areas is possible in theory, political realities suggest that in practice this is likely to be a rare event. This is reinforced by the Commonwealth government policy of not allowing mining in national parks (Australia 1990). The Commonwealth government's recent *One Nation* statement indicated a softening of this approach with an undertaking to conduct assessments of the economic value and potential of an area prior to any land use decision, such as declaration of national parks (Keating 1992). However, given that the State governments, rather than the Commonwealth government, are responsible for the declaration and management of most national parks, such an initiative is likely to be primarily of symbolic value unless the States incorporate it into their decision-making processes.

Information Problems

A recurring theme in the land access debate in Australia has been the importance of information (Ecologically Sustainable Development Working Groups 1991, 138–49; Industry Commission 1991, 117). The ability to make sensible and efficient decisions on alternative land uses depends critically on the availability and quality of information about the costs and benefits of alternative land uses. In choosing between mining and other public land uses, the need for better information on environmental costs and benefits is matched by a need for information on mineral resource values. When the mineral potential of an area and the mining options for those resources are known, an objective decision can be made about whether development or conservation or some combination of the two should take place. However, current arrangements do not facilitate the generation of information, and little attention has been paid to the practical issues of information generation and the implications for land access policy.

Clearly, information is required about the range of attributes and values of areas that offer attractive alternative uses. This will encompass information about both the environmental attributes and conservation values of an area, as well as the mineral potential of the area and the mining options for those resources. On the environmental side, however, there is a lack of economic incentive for individuals to gather basic information on environmental resources. This is a result of the nonrival and nonexclusive nature of such information—there is a public-good element to that type of information. The development of environmental and ecological databases by government may partly assist in overcoming this deficiency.

This lack of incentive is in contrast to the incentive provided by the private-good nature of information on mineral potential of an area. The costs associated with exploration for minerals can be recouped through the sale of that information or through the development of a mine and the sale of mineral output (either on the explored site or elsewhere if the exploration is unsuccessful). However, the incentive for mining companies to explore and produce the necessary information on which to base decisions is significantly reduced under current arrangements, since there is a great deal of uncertainty as to whether a company will be able to capture the benefits from exploration in areas perceived to have high conservation values. For example, in a recent assessment of the environmental and mineral resource potential of the Shoalwater Bay area in Queensland, only very limited information was available on the mineral sands potential of the region (ABARE 1993a, 121–5). This severely constrains the information base on which objective decision making about resource management can proceed.

Filling this information gap is problematic. One approach that has been suggested is to make access to land for exploration somewhat easier than it currently is under the institutional arrangements in Australia, with the understanding that a decision on whether mining would be allowed to proceed would be made at some future stage (Industry Commission 1991, 119–20). Unfortunately, neither mining nor conservation interests find such a proposition attractive. Resistance by environmental groups to the extension of exploration rights is largely driven by the current presumption that if anything useful is found, mining will then inevitably take place irrespective of the economic viability of extraction. At the same time, this presumption is strongly held and advanced by the mining industry in order to minimise the risks associated with investment in exploration.

From a public policy perspective, both these views may be counterproductive in that they work against appropriate information becoming available to make objective decisions on alternative land uses. To argue that mining should always follow exploration or that exploration should never be allowed in the conservation estate because mining may follow is to ignore the potential for more valuable alternative uses of a location.

The concerns held on both sides may be seen, in large part, to reflect the absence of clear policy guidelines on multiple resource use, particularly the levels of environmental performance required for operating in areas of environmental sensitivity. Clearly the costs of appropriate environmental compliance in environmentally sensitive areas will be considerably higher than in areas of less environmental significance in order to protect the high environmental values of the area. Environmentalists react to the absence of appropriate guidelines with a concern that ad hoc policies, developed in response to particular conflicts over resource use, may be insufficiently stringent, thereby allowing approval of the development of resources that would be subeconomic when faced with appropriate environmental compliance costs.

Miners react to the absence of appropriate guidelines with concern that ad hoc policies developed as a result of particular conflicts may be excessively stringent, thereby preventing the development of economic resources. They are concerned that resources that could meet the cost of (appropriate) environmental compliance will be prevented from being developed. This uncertainty may lead to suboptimal levels of exploration information being available.

The use of clear, appropriate, and (preferably) tested policy guidelines for multiple resource use may assist in improving the likelihood of appropriate information being generated and in reducing the conflict over resource access. This would require that the process by which companies can move from exploration to development, as well as the environmental criteria that would apply at each stage, be clear, transparent, and predictable. Ultimately, some level of political acceptability will be required if the concept of opening up the conservation estate to low-impact exploration and periodic review of resource uses is to be embraced, given that mining may follow and may involve some loss of environmental amenity. An important requirement is widening the understanding of the expanding technical capabilities of the mining industry with respect to resource use compatibility, especially the ability of miners to minimise off-site effects and to rehabilitate mine sites.

Ethical Systems and Valuation

Another fundamental cause of conflict over access to land is the rival ethical systems used by the groups engaged in the conflict.[8] The fact that opposing views are held by different groups in society does not, in itself, represent an impasse in terms of ensuring that natural resources are allocated to their most appropriate use. It does, however, complicate public policy.

On one hand, some groups, including most mining development proponents, argue their case from an anthropocentric perspective. That is, the relationship between human beings and the natural environment is of prime importance, with the satisfaction that nature provides to humans being the basic measure of value. Values are expressed in terms of individuals' willingness to pay for environmental goods and services and their willingness to be compensated for reductions in those goods and services. On the other hand, conservation groups tend to operate from a 'naturalistic' ethic, which seeks to extend the obligations of human beings to nature. They argue that economic value measures that arise from the expression of private preferences may be inappropriate as the sole value measures for public resource allocation.

From this position, the steps toward conflict between mining and some conservation groups over access to land are obvious. The conservation groups argue in terms of the 'sanctity of nature' and a belief in the rights of wildlife or nature in general, while the mining industry may emphasise the economic significance of development in terms of income, employment, and the trade balance. Conservationists argue against extending access to land for exploration because it is viewed as 'the thin edge of the wedge' and mining is seen as being bound to follow with more substantial environmental impacts. The mining industry, however, sees unfettered access to land as the lifeblood of the industry and economy. The conservation arguments have a strong flavour of concerns over the apparent irreversibility of mining development and the requirement for society to make a discrete choice between preservation of the environment and development.

These ethical considerations have been a major difficulty for Australian governments in assessing alternative uses of natural

---

[8]Kneese and Schulze (1985) provide a survey of the key issues in environmental ethics and economics.

resources. It could be argued that the political system may be the appropriate venue for addressing decisions that have a significant moral or ethical dimension. Attempts can be made, however, to incorporate these concerns into an objective framework for analysing alternative policy responses. Militating against the first approach is the fact that society now has in place political institutions and administrative processes that require conscious decisions to be made regarding the best use of scarce natural resources. This requirement implies that solutions will increasingly need to be reached using a framework for public decision making that is at least partly acceptable to all interests.

Fundamental to any framework is the concept of valuation. Mineral resources are produced and consumed within market structures that provide values for the minerals. Many environmental goods and services (such as aesthetic appeal and existence values), however, have no such market values. As already noted, this is primarily a result of the lack of property rights for these aspects of the environment. Assessment of the value of these intangible and qualitative benefits is a formidable task. The existence of rival ethical systems complicates but does not necessarily invalidate the decision-making process involved in resolving resource access conflicts. There is a well-established and increasingly accepted literature on valuation of environmental goods and services, which can go a long way toward redressing or reducing ideological conflicts.[9] In many ways, the argument that values cannot be placed on nature is an argument for humanity to do absolutely nothing, or to ignore the consequences of its actions on nature except to the extent that humanity is directly affected. Assigning values to aspects or components of the environment in ways that reflect the views and preferences of humans, however imperfect, is the only mechanism whereby decisions can be made as to whether, on balance, a human action is worthwhile. How this issue was addressed in the Coronation Hill inquiry is discussed below.

Irreversibility

The strong concern that many people feel for the fate of the environment presumably reflects a perception that some loss in environmental values will be irreversible. Many of the biological impacts of develop-

---

[9]See, for example, Randall 1986; Johansson 1987; and Pearce and Turner 1990.

ment activities can be extremely difficult to reverse over any measure of time that has any meaning for human societies. This may be seen as providing a basis for social decisions that are more than usually careful if environmental modifications are technically irreversible or if restoration of an environment to its original state is excessively costly in terms of resources or time.[10]

To what extent do concerns over irreversibility legitimately arise in relation to exploration and mining? As noted above, most exploration that is undertaken with modern techniques can have a negligible impact on the environment. While mining itself can have a more significant environmental impact, the use of modern mine management and rehabilitation techniques can significantly reduce the long-term impacts. In this regard, an important economic issue arises in determining the extent to which rehabilitation should be undertaken (Barnes and Cox 1992). Complete environmental restoration of a mine site may not necessarily be either technologically feasible or economically efficient, in that the costs of restoration may exceed the benefits. Arriving at a definition of successful rehabilitation involves combining economic and ecological criteria so that the optimal level of ecosystem function can be identified. In some (or even most) cases, the resulting optimum will mean some change or reduction in the physical characteristics of an area. To this extent, mining could be said to have some irreversible consequences. Whether or not this constitutes a cause for policy concern is open to debate.

However, this does not solve an issue of concern to many advocates of more ecocentric ideologies—that of the loss of intrinsic values associated with the environment, particularly wilderness areas. Several areas of high geological prospectivity in Australia are in regions that have significant wilderness qualities. These regions include Kakadu, Cape York Peninsula and southwest Tasmania. Certainly, the development of mines in such areas will reduce the wilderness values that some people may attach to the areas. The policy issue facing decision makers is the extent to which it is necessary to incorporate loss of existence values into the decision-making framework. In the final analysis, this may be seen to be a question of political judgement on the social value of

---

[10]There is a substantial literature on the role of irreversibility in the approach society should take towards economic development and environmental preservation. See, for example, Krutilla 1967; Arrow and Fisher 1974; and Krutilla and Fisher 1985.

wilderness. However, as is discussed below, there is good evidence that large sections of the Australian community place relatively high values on wilderness, although the extent to which the policy framework currently acknowledges or addresses the valuations is limited.

Benefit-Cost Analysis and Coronation Hill

The fact that society now has in place institutions and administrative processes that require conscious decisions to be made regarding the 'best' use of scarce natural resources implies that solutions will increasingly require some acceptable framework for public decision making. In principle, partial or complete trade-offs need to be made between alternative uses. Benefit-cost analysis provides one framework that allows the consideration of such trade-offs and has formed the basis of environmental policy development in Australia in recent years. However, the extent to which it has been formally applied has varied markedly among projects, industries, and levels of government and in the level of analytical sophistication. On only rare occasions are decisions made on the basis of, or with the assistance of, full social benefit-cost analyses. The experience of the Coronation Hill inquiry provides a useful case study of the application of a benefit-cost framework in one particular case and the challenges faced by governments in coming to grips with the strengths and weaknesses of the approach.

The basics of benefit-cost analysis are conceptually straightforward.[11] The simple question to be answered is whether the costs of a particular course of action exceed the benefits from that course of action, relative to alternatives (all measured at a particular point in time). The basic decision-making concept underlying benefit-cost analysis is that if, for some course of action, there is a potential excess of benefits over costs, it would (at least in theory) be possible to fully compensate all losers and still have a net social gain. Redistribution of the benefits from gainers to losers is not actually required.

Similarly, the technique of benefit-cost analysis is well known and conceptually simple. First, the full set of possible actions and the con-

---

[11]There is a large and growing literature on the theory and application of benefit-cost techniques, and it is beyond the scope of this lecture to fully examine the debate surrounding their use. For surveys of the range of issues involved, see Mishan 1982 and Dixon et al. 1988.

sequences of each action must be identified. Next, values must be assigned to each of these consequences, whether positive or negative. Those values must be reduced to a common denominator, usually financial, taking account of differences in the value over time. Finally, all the positive and negative values are summed. If a net positive value emerges, then the action is said to provide society with a net gain in welfare.

Estimating the benefits and costs of policy changes or actions is complicated by the fact that there are often uncertainties attaching to the possible outcomes. Uncertainty often exists, for example, about the future course of mineral prices, the extent and quality of particular mineral deposits, and the environmental costs of mining at particular sites. One method of overcoming uncertainty in the benefit-cost framework is to postulate a number of scenarios and observe their impacts on the net benefit or cost. An alternative, and often preferable, approach is to estimate a probability of occurrence of the potential events or impacts and take this together with the values of the outcomes to obtain an expected value. The expected values of the benefits of some changes are then compared to the expected values of the costs.

There are a number of other well-known practical difficulties with applying benefit-cost analysis. Chief among these is the difficulty of attaching values to environmental assets or services that may be threatened but for which markets do not exist or are imperfect. In the case of mining, effective markets do exist for the outputs of the industry and for the capital and labour inputs to mining, making valuation of that part of the consequences of mining relatively straightforward. This is not the case for many of the conservation services that might be affected if mining were to proceed in an area. Mining may involve opportunity cost in terms of conservation services forgone. As there is usually no market for many of these conservation services, it is difficult to determine their value in monetary terms.

During the 1980s, these nonmarket aspects of resource use have been considered in greater depth, and significant efforts have been made to develop techniques to estimate the value of nonmarket goods and services. These techniques vary from indirect methods (such as the use of market data for related goods) to direct methods (such as community attitude surveys and contingent valuation surveys). Valuing the off-site benefits from conservation of particular areas is the type of problem for which contingent valuation techniques have sometimes been used. Contingent valuations are based on directly questioning respondents to a survey that is designed to determine the willingness of people to pay for a good or to accept compensation for its loss, contingent upon some description of the good or scenario and the possible change in its availability or level of amenity.[12] Contingent valuation has been extensively used in other countries, particularly in the United States, where it has been the subject of considerable debate and refinement and has gained a degree of acceptance in court cases involving environmental damages. There have also been limited applications in Australia, and Wilks (1990) and Young (1991) provide surveys of the Australian experience in the use of nonmarket valuation techniques. The use of contingent valuation in the Coronation Hill inquiry was the first mining-related application in Australia.

The Coronation Hill Experience

Quantitative analysis of the conflict between mining and conservation at Coronation Hill was conducted on two fronts. First, an analysis was undertaken of the direct economic costs and benefits arising from the proposed Coronation Hill mine. A second line of analysis examined the values attaching to conservation of the area and sought to place monetary estimates on the nonmarket benefits.

The Mining Option. In the first of these analyses, a traditional benefit-cost study was undertaken using an indicative cash flow model of the planned mine to assess the economic worth of the proposed mine (ABARE 1990). This provided an estimate of the net present value to the Australian community of the mine should it be developed. It measured the economic surplus generated by the project, calculated as the future income stream from the sale of mine output, net of operating costs, and the opportunity cost of capital. The analysis indicated the mine would provide a net present value to the Australian community of about Aus$82 million in 1991 values.

The costs of preventing undesirable harm to the environment can be included in such an evaluation to the extent that such costs can be identified and measured. For example, if the cost of backfilling the pit and carrying out additional environmental monitoring were to be

---

[12]A detailed history, rationale, and practical guide to the technique are provided in Mitchell and Carson 1989.

included, the net national benefits would be reduced to around Aus$64 million. Had significant off-site environmental damage been identified, that too could have been brought to account in terms of the cost involved in preventing or rectifying such damage.

Clearly, there are limitations to the precision of this measure. First, there is a degree of uncertainty about the future course of many variables underlying the benefit-cost analysis, such as gold prices, the life of the mine, and so on. The use of scenarios that postulate alternative paths for these variables and observe their impact on the net benefit or cost is unlikely to be particularly satisfactory in overcoming this uncertainty. Decision makers can be left with a range of scenarios and little idea of the likelihood of the scenarios occurring. Policy development would be better served by attaching probabilities to impacts and outcomes and arriving at expected values for net benefits or costs.

Second, the choice of a discount rate to reduce the future benefits and costs to current dollars has generated much debate. Some groups, most notably those representing environmental interests, have argued against such discounting, claiming that current and future generations should be treated equally.[13] However, not discounting does not necessarily achieve such an outcome. Using a zero discount rate would, in principle, rationally imply reducing current consumption to subsistence levels in order to reap the gains in future consumption generated by the compound interest on the savings. The coexistence of positive rates of interest and consumption levels well above subsistence indicates that society does indeed have a positive discount rate, though what that rate should be is difficult to agree upon in practice. The appropriate discount rate for use in benefit-cost analysis is generally agreed to be the social time preference rate. However, given that risk markets are incomplete, that governments have some opportunity for pooling and spreading risks, and that individuals are generally risk averse, the possibility exists that the private discount rate exceeds the social time preference rate. If this is the case, all other things being equal, resources would tend to be exploited at a rate greater than is socially optimal.

What Price Preservation? The second avenue of quantitative analysis attempted to assess the preservation value of the Kakadu

---

[13]See Pearce and Turner 1990, 211–25, for a useful review of environmental concerns about discount rates.

Conservation Zone by undertaking a contingent valuation study of the zone. The purpose of the contingent valuation study was to estimate the value that people place on the preservation of the Conservation Zone in its current state, that is, without mining proceeding (Imber, Stevenson, and Wilks 1991). The preservation value included both use and nonuse values and so included future use and existence values. In the survey, 2,304 people were interviewed nationally, and a parallel survey of 502 people was conducted in the Northern Territory. In an attempt to encompass the diverse schools of thought about the likely impact of the mine, the survey was divided into two scenarios: one representing the case of major environmental damage if the mine went ahead, and the other couched in terms of minor environmental impact. Unfortunately, the survey was completed before further research on the likely environmental impact of the proposed mine became available to the RAC. This research indicated that a properly managed mine would have only a limited impact on the region's environment.

The results of the survey indicated that Australians felt strongly about potential environmental damage from mining in the Conservation Zone (Imber, Stevenson, and Wilks 1991, ix). The national survey showed that the median estimate that the people surveyed were willing to pay to prevent the mine was Aus$123.80 a person a year for ten years under the major impact scenario, and Aus$52.80 a person a year for ten years under the minor impact scenario. Given the minor environmental impacts likely to result if the mine went ahead, most attention was paid to the minor impact scenario. This estimate amounts to a national total of nearly Aus$650 million a year for ten years, a stream of payments with a present value of around Aus$4.3 billion. For the Northern Territory sample, the willingness to pay figures were substantially smaller, with Aus$7.40 for the major impact scenario and Aus$14.50 for the minor impact scenario.

The fact that the estimated value attached to preservation of the Kakadu Conservation Zone dwarfed the estimated net economic benefits from the mine proceeding caused a great deal of consternation and controversy among industry and government analysts. It could be argued that if the benefit estimate were really that large, a discussion over the issue of mining in the Conservation Zone would never be taking place. Even if there were severe failures of the political system, it seems unlikely that projects with a benefit-cost ratio of around 55:1 could be rejected or seriously questioned.

The RAC study, therefore, was subjected to a great deal of scrutiny, and there was considerable debate about the theory and practice of contingent valuation (see Resource Assessment Commission 1991b for some of this debate). The public controversy was defused by the RAC dropping any reference to contingent valuation in its final report, which concluded that it was not possible to estimate the true value of preservation in this case. Such a finding is more a reflection of the relative inexperience of the Australian policy process in the use of the technique and concerns over some of the technical issues involved than an outright rejection of the technique or its application. The number of contingent valuation studies conducted in Australia is relatively small but growing. However, in the studies that have been conducted, the magnitudes involved at the aggregate level appear substantial and reflect the significance placed on environmental resources by the community. For example, a study on the Great Barrier Reef reports an aggregate willingness to pay of around Aus$50 million a year for reef management, while the aggregate willingness to pay for forest preservation on Fraser Island is substantially higher at Aus$645 million a year (Young 1991, table 4).

Insights for Policymakers. If economic development and environmental protection were the only issues, there was not a great deal at stake in the government's decision on Coronation Hill, at least in substantive terms. The economic benefits were not so large, nor was the environmental impact very great. On the mining side, it was clear from analysis of the direct economic benefits and costs that the project would represent an efficient use of resources. But clearly it would be no economic bonanza. The mine would contribute substantially to the Northern Territory economy, especially in the construction stage, and would improve Australia's trade balance.

On the environmental front, the RAC found that the Conservation Zone was closely linked with the Kakadu National Park, especially through the South Alligator River, which serves as a refuge and a corridor for terrestrial and aquatic fauna. Mining development might detract from the ecological integrity of the combined Zone and National Park area, but the RAC concluded that a single mine, properly managed and monitored, would have a small and geographically limited direct impact on the known biological resources of the Conservation Zone (Resource Assessment Commission 1991a, xxi). There may, however, be some reduction in the existence values attached to the Kakadu region as evidenced by the results of the contingent valuation survey.

In summary, the Coronation Hill inquiry achieved what the RAC process was designed to achieve. All available information and research were assembled, and the options for the Commonwealth government were clarified. Indeed, the information set that was made available to government was one of the largest and most comprehensive yet produced for a single mining venture in Australia. However, the inquiry illustrated the difficulty in developing a policy framework in which the competing values of all factors can be effectively weighed. A simple comparison of direct costs and benefits does not always allow development proposals to be considered within a framework that captures the full range of values placed on environmental assets or services. A wider valuation framework appears to be needed to allow those values that accrue off-site, particularly the existence values held by the wider community, to be explicitly included in the decision-making process.

At the same time, the considerable scepticism that greeted the RAC contingent valuation survey indicates that Australia still has some way to go before techniques for valuing nonmarket benefits are accepted as part of the decision-making framework. Issues that remain to be resolved, and which lie at the heart of the contingent valuation debate in Australia, include the accuracy of responses in a hypothetical market framework, framing bias, the problem of aggregation, and the relationship between attitudes as reflected in contingent valuation responses and actual behaviour (ABARE 1991). Many of these problems have been encountered elsewhere in the world in contingent valuation studies, and methods to address them have been the subject of much research. The assessment by Hoehn and Randall (1987) that contingent valuation represents a program in progress, rather than the final word in environmental valuation, seems eminently appropriate in the Australian context. However, it is useful to briefly review some of the major concerns with the technique, concerns which have arisen in relation to resource access policy in Australia.

First, it is not clear that in the Kakadu application the method provides plausible estimates of monetary value for goods and services that people are not accustomed to valuing or paying for directly. The validity of a contingent valuation rests on the correctness of the assumption that all people are willing and able to place a monetary value on the goods, or management changes, in question. There are many things which at least some people may not think of in terms of monetary value. Sagoff (1988) argues, for example, that people do not think of many environmental questions in terms of individual market choices and money values.

Second, this problem is likely to be exacerbated when a controversial and politically sensitive issue, such as mining in Kakadu, is involved and when the survey respondents are abstracted from the process by which decisions are actually made about the goods. Carrying out contingent valuations of goods of potential national significance, such as Kakadu National Park, is costly. In the case of the Coronation Hill contingent valuation survey, many respondents apparently responded with opinions about desirable policy, rather than with money values. Responses tended to be polarised into responses that demonstrated support or opposition to a central proposition.

Further, national issues such as Kakadu are likely to be more problematical in terms of determining the appropriate population over which to aggregate individual willingness to pay estimates than are regional issues, where the goods and services in question are familiar to respondents. The choice of the entire adult population of Australia may be appropriate, given that Kakadu is one of Australia's best-known national parks. However, it may be that the use of contingent valuation in the study of an issue that has less political sensitivity would allow a more objective evaluation of the worth of the technique in the decision-making process. A smaller regional issue may not have the drawbacks associated with political responses and concerns over the appropriate level of aggregation.

Third, concerns over the extent of the market mean that the survey must be very carefully framed to avoid bias and to focus clearly on the good in question as distinct from other related goods. For example, the damage scenarios in the Coronation Hill survey were developed before important information on the off-site risks of the mine was available, so that the scenarios were more pessimistic and less clear with regard to the extent of environmental risk than they would have been if such information had been available. As a result, some respondents in the survey may have been valuing the whole of Kakadu National Park, rather than just the environmental assets that may have been placed at risk from mining in the Conservation Zone. The off-site risks of the mine were subsequently estimated to be relatively minor. It is unfortunate that the survey had to be conducted at the beginning rather than at the more informed end of the inquiry process.

Finally, if such valuation of nonmarket benefits is rejected, many decisions concerning possible environmental damage will be required to be resolved by protracted public debate and eventually by the political process. Moreover, estimates of the monetary valuation of environmental assets are likely to become part of that debate, even if only introduced to influence the political process. There are obvious advantages if the debate and the institutional responses can be developed in a common decision-making framework.

The Coronation Hill experience also highlights the problems that may arise when attempting to ensure that relevant data on the mineral potential of areas of apparent high conservation value are generated and made available to decision makers. Driving a wedge between the decision to allow exploration and the decision to allow mining will not necessarily result in private companies generating these data if there is no clearly defined, transparent, and consistent process in place by which companies would be allowed to move from exploration to mining development. The decision to set aside the Conservation Zone was ostensibly made to allow the mineral potential of the area to be assessed. However, no further exploration took place in the Conservation Zone, even though other sites were at least as (or, in the opinion of some, more) prospective than Coronation Hill (for example, at El Sherana). The exploration data on which the Coronation Hill mine plan was based had been gathered before Stage 3 of the Kakadu National Park and the Conservation Zone was declared. No further information on the mining potential of the area was generated because the risks associated with government decision making were perceived to be too high by mining companies.

The key lesson for land access policy is that the decision-making process should provide clear and consistent guidelines covering expectations about environmental performance (among other things), which potential exploration and mining companies need to know in advance if they are to make their own evaluations of whether to proceed. Reducing as far as possible the scope for ad hoc decision making will assist reducing the uncertainty the mining industry faces over access to land for exploration and mining.

Conclusion

Resolution of land access issues between mining and conservation groups is unlikely to ever be easy. This was highlighted by the debate over Coronation Hill. Policymakers were forced to address many of the fundamental issues underlying land access conflicts, particularly those dealing with defining significant environmental impacts, nonmarket values for environmental goods and services, and the overall role of economic analysis in resource-use decision making. Although the government decision to ban mining was disappointing to the mining industry, there are a number of lessons that can be learned from the process, lessons which are valuable for dealing with future conflicts. First, the inquiry found that mining and conservation can be compatible, provided appropriate environmental safeguards are in place. As a result, both the mining industry and decision makers can be more confident that an open and informed assessment of multiple resource use has the capacity to provide outcomes that may maximise social welfare. Second, the concern of Australian society about the existence values associated with areas of environmental significance should not be underestimated. Nor, indeed, should they be overestimated. Technical problems associated with measuring these nonmarket values need to be further investigated and the techniques further refined in order that decision makers have sufficient information with which to perform their tasks. Finally, a recurring theme throughout this lecture has been the importance of information for objective decision making. Institutional structures need to allow an efficient amount of information to be generated and used in an open and transparent framework, so that the intensity and intractability of conflicts can be reduced over time.

References

ABARE (Australian Bureau of Agricultural and Resource Economics). 1990. *Mining and the Environment: Resource Use in the Kakadu Conservation Zone.* Submission to the Resource Assessment Commission. Canberra: Australian Government Publishing Service.

———. 1991. *Valuing Conservation in the Kakadu Conservation Zone.* Submission to the Resource Assessment Commission. Canberra: Australian Government Publishing Service.

———. 1993a. *Shoalwater Bay Military Training Area Resource Assessment.* Canberra: Australian Government Publishing Service.

———. 1993b. *Commodity Statistical Bulletin.* Canberra: Australian Government Publishing Service.

Arrow, K. J., and A. C. Fisher. 1974. Environmental preservation, uncertainty, and irreversibility. *Quarterly Journal of Economics* 88 (2): 312–9.

Australia, Commonwealth of. 1990. *Ecologically Sustainable Development: A Commonwealth Discussion Paper.* Canberra: Australian Government Publishing Service.

Australian Mining Industry Council. 1990. *Shrinking Australia: Australia's Economic Future: Access to Land.* Canberra: Australian Mining Industry Council.

Barnes, P., and A. Cox. 1992. Mine rehabilitation: An economic perspective on a technical activity. In *Rehabilitate Victoria: Advances in Mine Environmental Planning and Rehabilitation.* Proceedings of The Australian Institute of Mining and Metallurgy Conference. Publication Series No. 11/92, 149–158.

Bolton, G. 1981. *Spoils and Spoilers: Australians Make Their Environment, 1788–1980.* Sydney: George Allen and Unwin.

Dixon, J. A., R. A. Carpenter, L. A. Fallon, P. B. Sherman, and S. Manipomoke. 1988. *Economic Analysis of the Environmental Impacts of Development Projects.* London: Earthscan Publications.

Ecologically Sustainable Development Working Groups. 1991. *Final Report: Mining.* Canberra: Australian Government Publishing Service.

Eggert, R. G. 1989. Exploration and access to public lands. *Resources Policy* 15 (2): 115–30.

Galligan, B., and G. Lynch. 1992. *Integrating Conservation and Development: Australia's Resource Assessment Commission and the Testing Case of Coronation Hill.* Discussion Paper No. 14. Canberra: Federalism Research Centre.

Grove, N. 1988. Quietly conserving nature. *National Geographic* 174 (6): 818–44.

Hoehn, J. P., and A. Randall. 1987. A satisfactory benefit cost indicator from contingent valuation. *Journal of Environmental Economics and Management* 14: 226–47.

Imber, D., G. Stevenson, and L. Wilks. 1991. *A Contingent Valuation Survey of the Kakadu Conservation Zone.* Resource Assessment Commission Research Paper No. 3. Canberra: Australian Government Publishing Service.

Industry Commission. 1991. *Mining and Minerals Processing in Australia, Vol. 3: Issues in Detail.* Canberra: Australian Government Publishing Service.

Johansson, P-O. 1987. *The Economic Theory and Measurement of Environmental Benefits*. Cambridge: Cambridge University Press.

Johnson, C. J. 1990. Ranking countries for minerals exploration. *Natural Resources Forum* 14 (3): 178–86.

Keating, P. J. 1992. *One Nation*. Canberra: Australian Government Publishing Service.

Kneese, A. V., and W. D. Schulze. 1985. Ethics and environmental economics. In *Handbook of Natural Resource and Environmental Economics*, vol. 1, edited by A. V. Kneese and J. L. Sweeney, 191–220. Amsterdam: North-Holland.

Krockenberger, M. 1992. *Mining: The Conservation Movement Agenda*. Paper presented at AIC Mining and Environment Conference, 2–3 July, Sydney, Australia.

Krutilla, J. V. 1967. Conservation reconsidered. *American Economic Review* 57: 777–86.

Krutilla, J. V., and A. C. Fisher. 1985. *The Economics of Natural Environments: Studies in the Valuation of Commodity and Amenity Resources*. Washington, D.C.: Resources for the Future.

Mishan, E. J. 1982. *Cost Benefit Analysis: An Informal Introduction*. 3d ed. London: George Allen and Unwin.

Mitchell, R., and R. Carson. 1989. *Using Surveys to Value Public Goods: The Contingent Valuation Method*. Washington, D.C.: Resources for the Future.

Pearce, D. W., and R. K. Turner. 1990. *Economics of Natural Resources and the Environment*. New York: Harvester Wheatsheaf.

Randall, A. 1986. Valuation in a policy context. In *Natural Resource Economics: Policy Problems and Contemporary Analysis*, edited by D. W. Bromley, 163–99. Boston: Kluwer Nijhoff Publishing.

Resource Assessment Commission. 1991a. *Kakadu Conservation Zone Inquiry: Final Report*. Canberra: Australian Government Publishing Service.

———. 1991b. *Commentaries on the Resource Assessment Commission's Contingent Valuation Survey of the Kakadu Conservation Zone*. Canberra: Australian Government Publishing Service.

Sagoff, M. 1988. *The Economy of the Earth: Philosophy, Law, and the Environment*. Cambridge: Cambridge University Press.

Wilks, L. C. 1990. A *Survey of the Contingent Valuation Method*. Resource Assessment Commission Research Paper No. 2, Canberra: Australian Government Publishing Service.

Young, R. 1991. The economic significance of environmental resources: a review of the evidence. *Review of Marketing and Agricultural Economics* 59 (3): 229–54.

Zimmerman, D. 1992. Let's have economic impact statements to assess red tape's growing effects. *The Mining Review* 16 (1): 5–8.

Mining Waste and the Polluter-Pays Principle in the United States

JOHN E. TILTON

Mining is essential for modern civilization. Without it, there would be no automobiles or televisions, no telephones or x-ray machines, no skyscrapers or computers.

Mining also disturbs the environment. The production of a ton of copper can generate five hundred tons of waste, depending on the grade of the ore and the amount of overburden removed.[1] The resulting waste piles, tailings ponds, and slag dumps are unsightly. The leaching of cadmium, arsenic, and heavy metals from surface and subterranean workings contaminates streams and rivers as well as underground

This lecture was written while the author was a Fulbright Research Scholar with the Centre d'Economie Industrielle at the Ecole Nationale Superieure des Mines de Paris. It is based on a larger study currently being conducted at the Colorado School of Mines on Superfund and the cleaning up of old mining sites.

The support, financial and in kind, provided by the John M. Olin Foundation, the Viola Vestal Coulter Foundation, Resources for the Future, and the Ecole des Mines de Paris is gratefully acknowledged. Without implicating, I would also like to thank Roderick G. Eggert, Wade E. Martin, and Katherine N. Probst for their helpful comments on earlier versions of this study, and Kathleen Anderson for her enthusiastic assistance and support in gathering and analyzing information.

[1]According to Ayres (1992, Table 1), the U.S. Bureau of Mines estimates that a ton of copper requires the handling of 337 tons of material if it is produced from an underground mine and 550 tons if it is produced from a surface mine.

aquifers. Windblown dust deposits lead and other toxic materials on the surrounding area, poisoning wildlife, livestock, and even people.

For years public policy, particularly at the federal level, largely ignored the environmental costs of mining and mineral processing. Most mining took place in remote and sparsely populated areas, and in any case the public was preoccupied with economic growth and other priorities. What little regulation there was, one found at the state and local levels.

As is well known, the environmental movement of the 1960s and 1970s, reflecting a radical reordering of public priorities with respect to the environment, abruptly changed this situation. Within a few years, Congress passed new laws and greatly strengthened old ones to give the federal government and its newly created Environmental Protection Agency (EPA) a major role in pollution control and abatement.

Within the mining community, the two federal statutes of possibly the greatest concern are the Comprehensive Environmental Response, Compensation, and Liability Act (CERCLA) of 1980—better known as Superfund—and the Resource Conservation and Recovery Act (RCRA) of 1976. Both pieces of legislation, along with subsequent amendments, deal broadly with hazardous substances and are not confined solely to mining and mineral-processing wastes. Within the mining sector, CERCLA has particular relevance for the cleaning up of old mines, RCRA for operating mines. Both purport to make the parties responsible for the pollution, where possible, pay for remediation in accordance with the widely accepted polluter-pays principle.

In this lecture, however, I argue that the polluter-pays principle is not applicable for cleaning up old mining sites and that current efforts under CERCLA to make responsible parties pay are actually counterproductive. While the polluter-pays principle is appropriate for the wastes associated with current mining, these wastes—for the time being at least—have been exempted from federal regulation under RCRA.

I am, however, getting somewhat ahead of my story. The important point for the moment is simply to make clear what this lecture does and does not address. Its purpose is to examine the proposed allocation of costs for cleaning up mining wastes under CERCLA and RCRA and to assess their efficacy.

There are, of course, many other important policy issues associated with these statutes—for example, the procedures for determining the nature and extent of the cleanup and the role of costs in that process—

but they are not addressed here. Moreover, my focus is solely on CERCLA and RCRA. The mining industry is subject to numerous other laws and regulations at both the federal and state levels. While these are important, they receive little or no attention here.

The analysis begins in the next section with an examination of the polluter-pays principle and the reasons for its nearly universal acceptance. The two subsequent sections then look respectively at CERCLA and old mining sites and at RCRA and current mining sites. Since, in my view, the shortcomings of CERCLA are the more serious, CERCLA receives much more attention. The final section highlights the findings and examines their implications.

The Polluter-Pays Principle

Mining and many other economic activities incur two types of costs—those the producing firm must pay, as is normally the case for labor, capital, and material inputs, and those the producing firm does not pay, as is often the case for water pollution and other forms of environmental damage. The latter costs, because they are external to the producing firm, are commonly referred to as *externalities*. While the firm and, ultimately, consumers of the final product may avoid the external costs, society as a whole is not so fortunate. Municipalities downstream from the polluting facility, for example, must spend more to treat their water or incur the costs of developing alternative sources of water.

As with all scarce resources, there is an optimum level of use for any given environmental resource. For society as a whole, this occurs at the point where the additional benefits society derives by permitting one more unit of pollution are just equal to the additional costs society incurs. In economic terms, the optimum level of use occurs where the marginal social benefits (MSB) equal the marginal social costs (MSC).[2]

In Figure 1, which is a fairly standard diagram found in many environmental and natural resource economics textbooks, this optimum level is at point P_0 The negative slope of the MSB curve reflects the assumption that goods and services with lower (net) social benefits per unit of pollution will be produced as the permitted level of pollution

---

[2]The social benefit of an additional unit of pollution is the net value to society of the goods and services that the additional unit of pollution makes possible.

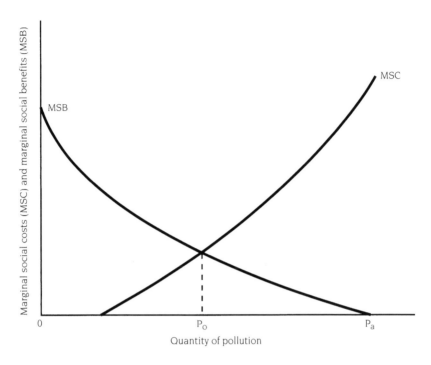

Figure 1. The marginal social costs and marginal social benefits of pollution.

increases. The positive slope of the MSC curve implies that as the level of pollution increases the social costs incurred for each additional unit of pollution rise. Both are reasonable assumptions.

Where the social costs of even small amounts of pollution are very high (or where the social benefits from even the first units of pollution are negligible), the MSC curve may lie entirely above the MSB curve, indicating an optimum pollution level of zero. This, however, is unusual. Normally, the optimum is positive, reflecting the fact that some pollution is socially desirable and that efforts to reduce pollution beyond some point will actually reduce the welfare of society.

Now if all the social costs and benefits of pollution are incurred or internalized by the producing firm, the firm will have an incentive to continue to pollute only up to point P_o, as beyond that point the additional cost it incurs exceeds the additional benefit. In this case, the market works well, producing the optimum without government intervention.

This is not the case, however, with externalities. If, for example, the producing firm realizes all the benefits associated with its pollution but none of the costs, it has an incentive to expand its production until the additional benefits from pollution are zero and pollution has reached point P_a in Figure 1, far beyond the optimum P_o.[3]

Economists have long been aware of the inefficiencies introduced by the externalities associated with environmental damage. When producers and ultimately consumers are not charged for the environmental resources used in producing goods and services, excessive pollution occurs for several reasons.

First, since consumers do not pay the full social costs of pollution, pollution-intensive goods are underpriced relative to other goods, and hence overconsumed and overproduced. In short, the economy fails to achieve *allocative efficiency*, in that resources are not being allocated in a way that produces the optimal mix of goods and services. Instead, there is a bias in favor of underpriced, pollution-intensive goods.

Second, producing firms, to the extent they can, will substitute environmental resources, which for them are free, for labor, capital, and other inputs, for which they must pay. This causes a breakdown in *production efficiency*. Production processes are employed that fail to minimize the costs to society of producing goods or services because they use environmental resources more intensively than is desirable.

While both allocative and production efficiency are important, over the longer run failure to achieve *dynamic efficiency* is far more critical. If producing firms are charged the full social costs for the environmental resources they consume, they have strong incentives to develop and adopt new environmental-saving technologies (that is,

---

[3]Under certain circumstances—well-defined property rights; few involved parties and hence negligible transaction costs; and the absence of strategic behavior—Coase (1960) has demonstrated that the market will produce the optimum level of pollution without government intervention. If, for example, the producing firm has the right to pollute a stream, once its level of pollution has exceeded the optimum P_o, the costs to downstream communities, dependent upon the stream for their municipal water supply, of additional pollution exceed the benefits to the producing firm. The downstream communities consequently have an incentive to pay the producing firm to restrict its pollution to P_o, and as long as the payment exceeds the additional benefits the producing firm would receive from further pollution, the firm has an incentive to agree. In most situations, however, the stringent assumptions of the Coase Theorem, particularly the assumption of negligible transaction costs, are not satisfied.

technologies that save environmental resources). Equipment manu-
facturers similarly are motivated to embody in their equipment new
processing technologies that pollute less per unit of output, because
this reduces their customers' costs and makes their equipment more
competitive.

However, if producing firms are not forced to pay for their pollution, neither they nor their suppliers have an incentive to develop and
adopt new environmental-saving technologies. One of the strongest
forces society has at its disposal for coping with resource scarcity is
effectively neutralized.

While economists and policy analysts espouse the polluter-pays
principle primarily for the efficiency considerations just examined,
public officials and environmental advocacy groups are attracted by
more political considerations. With growing public sentiment to minimize the role of government and particularly to reduce taxes, the polluter-pays principle solves the problem of where to find the funds to
clean up the environment.

The general public favors the polluter-pays principle largely for
equity reasons. From early childhood on, we all learn that if we make a
mess, we have to clean it up. It thus is only fair that those firms that
pollute the environment—and realize the benefits from doing so—
should pay to fix it.[4]

Given these attributes, the widespread acceptance that the polluter-pays principle enjoys is not surprising. Indeed, over twenty years
ago, the Organization for Economic Co-operation and Development
(OECD) stipulated that the polluter-pays principle should be a fundamental tenet of the environmental policies of its member countries,
most of the world's more developed nations.

Old Mining Sites and CERCLA

The number of old mining sites within the United States, though not
known for certain, is large, probably somewhere between 100,000 and

---

[4]As Portney (1991) has noted, equity in allocating the costs of cleanup can encompass
other considerations than the amount of pollution one has contributed. One might, for
example, want to take into account who benefits from cleaning up and the ability of different parties to pay. Which considerations equity should encompass and how to weight
them are value judgments over which reasonable people can differ.

400,000 sites.[5] Of these, several thousand may pose significant threats to public health and the environment. At the federal level, environmental policy has relied, for the most part, on CERCLA for cleaning up the more hazardous of the old mining sites.

Superfund Legislation

CERCLA, passed in the final days of 1980, reflects a shift in environmental policy, which during the 1970s emphasized air and water pollution, towards solid and particularly hazardous wastes. Contributing to this shift was the widely publicized Love Canal disaster: in 1978, homeowners in a middle-class residential subdivision in Niagara Falls, New York, found old chemical wastes, buried years earlier in the nearby Love Canal, seeping through foundation walls and into their homes. The impact of this disaster on public opinion was tremendous, and Congress wrote CERCLA with such relatively small, highly toxic, chemical waste sites in mind.

CERCLA requires EPA, with the assistance of the states, to identify existing sites hazardous to public health and the environment and to ensure the sites' proper remediation. By focusing on environmental damage created by past economic activities rather than current and future activities, CERCLA differs from most other environmental legislation.

To fulfill its statutory obligations, EPA has developed the Comprehensive Environmental Response Compensation and Liability Information System (CERCLIS), which contains information on over 30,000 potentially hazardous sites that have been brought to EPA's attention. EPA ranks sites according to its hazard-ranking system,[6] and the most threatening sites are placed on the National Priorities List (NPL). As of

---

[5]Robert E. Walline, a mining waste expert for the Environmental Protection Agency, estimates there are 200,000 old mining sites in the United States (personal communication). Philip M. Hocker, president of the Mineral Policy Center, suggests the number may be 400,000 (personal communication). The Western Interstate Energy Board (1991) has collected data from a number of states suggesting the number is somewhere between 100,000 and 200,000 sites, though the report strongly warns that the figures it reports for different states are not compatible and should not be aggregated.

[6]The EPA hazard-ranking system uses a number of criteria, such as the toxicity of the material contained on the site and the likelihood of human exposure, to assess each site. Using numerical weights for each criterion, EPA calculates a hazardous index, which is used in determining which sites are placed on the National Priorities List.

June 1990, the NPL contained about 1,200 sites, of which sixty were mining sites. An additional 220 mining sites are listed by CERCLIS.[7]

Of particular importance for our purposes, Congress in passing CERCLA assumed that the cost of cleaning up most Superfund sites would be borne by the responsible parties, and it provided EPA, as we shall see shortly, with some fairly powerful legal measures to ensure where possible that this would be the case. To pay for orphan sites for which responsible parties could no longer be found, Congress created the Hazardous Substance Trust Fund (from which comes the name, Superfund). EPA can also use the trust fund to initiate remediation at sites before the courts have determined the financial obligations of the responsible parties.

Initially, CERCLA authorized $1.6 billion for the trust fund over five years, to be provided from three principal sources: funds recovered from responsible parties; a special tax on petroleum and chemical feedstocks; and general tax revenues. In 1986, CERCLA was amended by the Superfund Amendments and Reauthorization Act (SARA), which increased the authorization for the trust fund to $8.5 billion over five years and provided an additional source of funds, an environmental tax on profitable corporations.[8]

As already noted, the law assumes that most of the cleanup will be funded by those responsible and defines the parties liable broadly to include the generators and transporters of hazardous waste, as well as the owners and operators of the sites. Individuals, small companies and large corporations, banks, and federal and state agencies, as well as local governments and schools, can all be potentially responsible parties (PRPs).

Liability is *strict*, so EPA need not demonstrate that PRPs were at fault by engaging in nonprudent, negligent or illegal behavior. Liability

---

[7]The data on mining sites are from EPA (Hoffman 1990). The Western Interstate Energy Board (1991) indicates that there are forty-eight mining sites on the NPL as of August 1991. Since its figures come from the same EPA source, the lower figure presumably excludes certain processing facilities, such as aluminum smelters, that are included in the sixty sites noted in the text.

While the number of mining sites on the NPL and CERCLIS is relatively small, some of the listed sites are quite large and individually encompass numerous old mines, mills, and smelters.

[8]This tax is 0.12 percent of a corporation's minimum taxable income over $2 million. See Dower 1990, 175.

is also *joint and several*, so EPA can hold a subset, or even one, of the PRPs responsible for the entire cost of cleaning up a site, regardless of how many other parties may also have been involved. (However, PRPs identified by the EPA can, in turn, sue other PRPs for compensation.) Finally, liability is *retroactive*, so EPA can require PRPs to pay for remediating sites that predate 1980 and CERCLA. Thus, the law gives EPA tremendous power to identify those parties responsible for hazardous waste sites and to force them to pay for their cleanup—in short, tremendous power to apply the polluter-pays principle.

Applicability of the Polluter-Pays Principle

Yet, a close look at how Superfund is actually working, at least for the mining sites I am investigating,[9] suggests that even with the best of intentions and the strongest legislation it is simply not possible to force those responsible to pay for remediation.

The reasons for this are well illustrated by the Sharon Steel Superfund site, located at Midvale, Utah, a small town some twelve miles south of Salt Lake City. Around 1906, the U.S. Smelting, Refining, and Mining Company began processing ores at Midvale, and in the mid-1920s the company built a modern flotation mill to process lead-zinc ores. U.S. Smelting also constructed a lead smelter at Midvale during this period.

Both before and during World War II, Utah was a major lead-producing region. Asarco also operated a lead smelter in the state, at Murray, and the International Smelting and Refining Company, a subsidiary of the Anaconda Company, had a lead mill and smelter at Tooele, Utah.

After the war, lead-zinc mining in Utah entered a long period of decline. U.S. consumption grew slowly, government stockpile purchases ceased in the mid-1950s, and competition from new mines in Missouri and overseas, particularly Peru, Canada, and Australia, was intense. Even with the closure of the Asarco smelter at Murray in 1949,

---

[9]My study is examining four Superfund mining sites: the Sharon Steel site at Midvale, Utah; the Clear Creek/Central City site in Colorado; the California Gulch site at Leadville, Colorado; and the Smuggler Mountain site at Aspen, Colorado. The last three sites were selected in part because they are in Colorado; the first site was selected because I was involved in this case as an expert witness and hence am familiar with it.

it was clear by the 1950s that the remaining mill and smelter capacity at Midvale and Tooele far exceeded that needed to treat the declining ore production of the region.

As a result, in 1958 U.S. Smelting entered into an agreement with International Smelting under which U.S. Smelting shut down its smelter at Midvale and shipped its concentrates to Tooele for processing, and International Smelting in turn shut down its mill at Tooele and shipped its ore to Midvale for milling. For fourteen years the two companies shared each other's facilities, but the lead market remained depressed, and after years of poor returns U.S. Smelting closed its Midvale mill in 1971 and formally terminated its agreement with International Smelting in 1972.

U.S. Smelting changed its name to UV Industries in the same year and extended its involvement in electrical equipment. It continued to have problems, however, and eventually went into receivership. Along the way—specifically in 1979, a year before the passage of CERCLA— U.S. Smelting sold the Midvale site to the Sharon Steel Company. By then, the mill as well as the smelter had been razed. Sharon Steel purchased the property for its own commercial purposes.

While the buildings were gone, the slag pile from the smelter and the tailings pond from the mill remained. The tailings pond alone encompassed some fourteen million tons of material in uncovered piles up to fifty feet deep covering some 260 acres. Resembling fine grain sand, this material blew off site. Some 44,000 people live within two miles of the site, and on occasion nearby residents have used the material for sandboxes and gardens. The tailings were first suspected of being a health problem in 1982, and subsequent tests confirmed the presence of potentially hazardous levels of lead, cadmium, arsenic, chromium, copper, and zinc in the soil, air, and groundwater. The site was proposed for the NPL in 1984. (The slag pile from the smelter was also determined to be a health hazard and added to the NPL at a later date.)

EPA identified and sued three PRPs—Sharon Steel, the current owner of the site; UV Industries Liquidating Trust, created to liquidate the remaining assets of UV Industries, the successor to U.S. Smelting; and the Atlantic Richfield Company (Arco), the large petroleum company that acquired Anaconda and its subsidiary, International Smelting, during the 1970s. In 1990 the three companies agreed to pay the government a total of $63 million, with each company contributing approximately a third of the total.

Just how much of the total cleanup costs the $63 million will cover is unknown at this time, since the site has been divided into two parts and the remediation plans for one part have yet to be determined. Nor do we know what proportion of the total tailings (or perhaps more appropriately, what proportion of the total hazardous materials) at the site was contributed by International Smelting during the fourteen years it had its ores milled at Midvale. What is clear, however, is that during the sixty-five years the Midvale mill was in operation, Sharon Steel contributed none of the tailings and hazardous waste, and International Smelting contributed far less than U.S. Smelting. Under the polluter-pays principle, Sharon Steel would not have been a responsible party.

Moreover, ultimately, it is people who are responsible for pollution, and it is people who pay to clean up the environment.[10] The managers and stockholders of Arco who paid for the remediation of the site in 1990 were quite a different set of individuals from those responsible for the tailings contributed by International Smelting at Midvale during the 1958–72 period. Similarly, the claimants to the remaining assets of UV Industries Liquidation Trust in 1990 were a very different group from those who owned and managed U.S. Smelting when it operated the mill. In short, the people who polluted were beyond the reach of the law and did not pay, while those within the reach of the law and who did pay were not the polluters.

Going one step further, one can argue that it was really the consumers of automobile batteries and other lead products over the 1906–71 period who were responsible for and benefited from the pollution at Midvale and other old lead mills. In most industries, the savings that firms realize when public policy fails to internalize the external costs of pollution are largely passed on to the consumer in the form of lower prices. This is clearly true in competitive industries, where price over the longer term is determined by the costs (that is, the internalized costs) of marginal producers. In industries where producers possess market power, it is also true to the extent that the producers take their costs into account in determining price.

---

[10]This does not mean that companies and corporations are not responsible for pollution, or that they cannot be made to pay for cleaning up the environment. The point is simply that corporate decisions are made by people, and that the costs firms incur in cleaning up the environment must be covered by higher prices to consumers, lower dividends to shareholders, lower salaries to employees, or some combination of these possible consequences compared to what they otherwise would have been.

Thus, even if the law could identify the managers and stockholders of U.S. Smelting and International Smelting over the period when these two companies deposited mill tailings at Midvale, a strong case can be made that not they but the millions of consumers of lead products between 1906 and 1971 were responsible for and ultimately benefited from the pollution at Midvale. Even if most of these individuals were still alive, which is probably not the case, they clearly would be beyond the reach of the current Superfund legislation, despite the broad powers it bestows upon EPA, and presumably beyond the reach of any legislation.

The problems in employing the polluter-pays principle are not unique to Midvale. They are found at all the old mining sites I am studying.[11] Indeed, EPA plans to finance the remediation of the Clear Creek/Central City site from the trust fund because of the difficulties in identifying and collecting from the numerous parties contributing mining wastes at the site over the last century.[12]

One might protest that this reasoning leads to the conclusion that corporations should never be held accountable for cleaning up their old hazardous sites, since by the time they are forced under CERCLA to undertake remediation, their stockholders and managers have to some extent changed. There is, however, an important distinction between the market information available today compared to the 1970s and earlier regarding firm liability for cleaning up hazardous sites. In 1979, when Sharon Steel purchased the Midvale site, it may have paid somewhat less for the site because the tailings piles made it more difficult to use parts of the site for particular purposes. It is difficult to argue, however, that Sharon Steel paid less for the site because the market foresaw the future liability the owner would incur for cleaning up the site. CERCLA had not yet been passed, and the concept of retroactive liability was quite unusual in the United States.

This, in turn, makes it difficult to make a convincing case that UV Industries actually covered Sharon Steel's share of the $63 million paid to clean up the site by accepting a lower sale price for the Midvale site,

---

[11]Similar difficulties in applying the polluter-pays principle apparently exist at nonmining Superfund sites as well. See, for example, Clay 1993 and Edelstein 1993.

[12]As noted below, Colorado has recently authorized limited gambling at Central City and the nearby town of Black Hawk, and EPA is considering naming the owners of new casinos built on old mining wastes as PRPs. Few, if any, of these PRPs, however, were responsible for the pollution.

and that U.S. Smelting, UV Industries' predecessor, anticipated this liability and thus considered the future costs of remediation in setting the prices it charged its customers.

Today, the situation is quite different. The market is much more sensitive to the potential liabilities of firms for cleaning up the environment under CERCLA. These liabilities are carefully considered during mergers and acquisitions and are much more likely to be reflected in the market value of firm equity. Firms, in turn, take these liabilities into account in their production and marketing decisions and are likely to provide products only if the consumer is willing to pay for all the firm's costs, including the costs it anticipates for cleaning up the environment. In this situation, the polluter-pays principle can be employed, and public policy should hold the producer accountable for its pollution to ensure that the principle is employed.

Effectiveness and Efficiency of Superfund

While Superfund is widely assumed to be based on the polluter-pays principle and, indeed, it is this belief that accounts for much of its public support, one might argue that it does not greatly matter whether or not the polluter actually pays, as long as Superfund is effective and relatively efficient in accomplishing its primary objective, the cleaning up of hazardous waste. Here it is useful to distinguish the impact and performance of Superfund in three areas—past behavior of producers and consumers, the remediation of old hazardous waste sites, and present and future behavior of producers and consumers.

Past Behavior. The production and consumption that have taken place in the past cannot now be altered. If environmental policy failed to force firms and consumers to internalize the full social costs of the environmental damage associated with their activities, neither Superfund nor any other legislation can retroactively correct the situation. This, of course, would be true even if those ultimately responsible and benefiting from pollution could now be identified and forced to pay.

The overconsumption of environmental resources in production by firms (production inefficiency) and the overconsumption of environmentally intensive goods by consumers (allocative inefficiency) cannot now be changed, any more than we can alter the rate and direction of past innovative activity to reflect a higher priority for new environmental-saving technologies (dynamic inefficiency).

To the extent that society is now burdened with the adverse consequences of such inefficiencies, the responsibility lies with past public policy and not with the behavior of firms and consumers furthering their own self-interests within the laws and conventions of their times. Until the passage of CERCLA, at least, our society asked individuals and firms to pursue their own welfare, constrained only by existing laws and regulations, not future laws whose content they had no way of foreseeing.

Moreover, even in judging the public policies of the past, we need to proceed with caution. As noted earlier, public interest in the environment increased dramatically during the 1960s and 1970s. The policies of the 1950s and before may have adequately reflected the environmental costs of mining and other activities, at least as perceived at the time. The country was less populated with far more open space and, in any case, had other higher priorities.

While most people would agree that every generation has responsibilities not only to its own members but to the members of subsequent generations, it is far from clear that the strong environmental concerns of the present generation could easily have been foreseen by earlier generations. With the passage of time, moreover, the concerns and priorities of earlier generations tend to fade from our memories, no matter how important the successful resolution of these concerns might have been for our own welfare. In this regard, it would be hard to condemn the generation of the 1940s, even if that generation had clearly seen the environmental storm clouds gathering on the horizon, for giving the defeat of Nazism higher priority than pollution.

Each generation passes on to the next many "goods," including its social institutions, knowledge base, and replenished human capital, as well as "bads," including its environmental damage. Overall, the generations of our fathers and grandmothers invested a sizable share of their current income, a greater share than the current generation, and likely left the country more improved for our generation than our daughters and grandsons will say of us, the billions of dollars we are investing in the environment notwithstanding. This is not to say that the environmental policies of the past were necessarily appropriate, but simply to note that condemning them is not as simple as some would suggest.

Remediation of Old Hazardous Waste Sites. If past behavior that damaged the environment cannot be undone, the damage itself can be addressed. Conceptually, we would like our current environmental policies to clean up the damage from past practices just to the point where the marginal social benefits equal the marginal social costs (see again Figure 1) and to do so in a timely manner that minimizes the total costs.

While Superfund was passed to remediate the damage from past practices, the following—all of which are associated with the preference to allocate the costs of remediation to PRPs—undermine its effectiveness and efficiency.[13]

Transaction Costs. These are the costs of determining the liability for cleaning up, as opposed to the actual costs of remediation. They cover the expenses of negotiation and litigation between EPA and PRPs, among various PRPs, between PRPs and their insurance companies, and between insurance companies and reinsurance companies; these expenses cover the costs of lawyers, expert witnesses, and independent site and damage evaluations (beyond those needed for remediation).

At the Sharon Steel site, the only Superfund site among those I am studying where the government has settled with the PRPs, the three responsible parties have, as noted earlier, paid $63 million for remediation. The technical costs and attorneys' fees incurred by EPA and the Department of Justice in negotiating and litigating this case are estimated at $3–$4.5 million.[14] Although similar figures are not available for Sharon Steel, UV Industries Liquidation Trust, and Arco, each probably spent as much—and perhaps twice this amount—suggesting that transaction costs accounted for somewhere between a fifth and a third of the total funds devoted to this site by the PRPs and EPA.

Just how typical the Sharon Steel site is in this regard is difficult to say. According to one industry official, a rule of thumb among lawyers and others is that 40 percent of all expenditures for Superfund sites are for legal expenses.[15] Acton and others (1992), in the only empirical study of which I am aware, found that during the 1984–89 period the transac-

---

[13]Given our focus on paying for remediation, only those factors arising from Superfund's tendency to identify PRPs are discussed. There are, however, other important factors that also reduce Superfund's effectiveness and efficiency, including the secondary role given to costs in determining the appropriate nature and level of remediation.

[14]The EPA Remedial Project Manager for the Sharon Steel site, Mr. Sam Vance, estimates the total EPA "response costs" at approximately $6.5 million, with between one-half and two-thirds of this amount going toward technical work and attorneys' fees in preparation for litigation (personal communication with Kathleen Anderson).

[15]Marcel F. DeGuire, Vice President, Environmental Affairs and Metallurgical R&D, Newmont Mining Corporation (personal communication).

tion costs incurred by each of five large industrial corporations averaged 21 percent of the total expenditures at all the Superfund sites at which these companies were involved as PRPs.[16] Acton and others also found that transaction costs varied greatly, accounting for a much higher percentage of total expenditures at sites with more than one PRP (such as the Sharon Steel site) and for sites in the early stages of cleanup.

The lack of comprehensive data on transaction costs makes it impossible to determine with any precision the magnitude of the expenditures devoted solely to determining who should pay for cleaning up Superfund sites. It is clear, however, these costs are far from negligible and represent real resources that presumably could be better spent on actual remediation or other social needs.

Property Values. Two of the Superfund sites I am examining (the California Gulch site at Leadville, Colorado, and the Smuggler Mountain site at Aspen, Colorado) contain residential developments where homeowners have suffered a substantial drop in property values and have found it difficult at times to sell their homes, in part the consequence of the potential liability any purchaser might face as a partial "owner" of a Superfund site. In addition, banks may understandably be reluctant to extend mortgages to interested buyers, since foreclosing on a delinquent borrower could result in the bank becoming a PRP.

Similarly, at the Central City/Clear Creek site, the threat of Superfund liability could dampen the economic expansion stimulated by the recent state referendum legalizing gambling at Central City and Black Hawk. The evidence to date, though, suggests the effect has not been great, since new casinos are being built on old mining waste piles.

Speed of Cleanup. Superfund is widely criticized for the time it takes to clean up hazardous sites, now estimated to average fifteen years.[17] As Probst and Portney (1992, 20–22) have pointed out, a number of factors contribute to the slow pace of remediation.

---

[16]The figure is much higher, between 80 and 90 percent, for the four insurance companies they studied.

[17]According to testimony by Jan Acton, Congressional Budget Office, at hearings before the Subcommittee on Investigations and Oversight of the House Committee on Public Works and Transportation in 1991, as cited in Probst and Portney 1992, 20, on average fifteen years may elapse between the time a site is brought to the attention of EPA and the completion of its remediation.

Among the sites I am studying, disagreements among EPA, the PRPs, and the local community regarding the most desirable approach to remediation have led to some extended delays. At the Sharon Steel site, for example, the city of Midvale objected to EPA's proposed solution because it would have precluded the use of the site for future commercial development. As a result, the planned remediation is now being reconsidered and may be changed completely. Similarly, at the Smuggler Mountain site in Aspen, local residents have questioned the basic models that EPA uses for assessing the health risk, challenging in the process the need for EPA's proposed remediation. The resulting standoff has deferred any cleanup for several years while a study is conducted on the lead absorbed by pigs after digesting tailings similar to those found on site.

The liability provisions of Superfund constitute another factor that many believe contributes to the slow pace of cleanup. While the influence of the liability provisions cannot be isolated from that of other factors, the time needed to identify PRPs, to conduct the necessary studies for litigation, and to carry out the required negotiations and legal actions is often substantial.

Remining. Over time, as new technology makes it possible to exploit lower grade and poorer quality ores, mining companies find it worthwhile to reprocess the waste piles found at many old mining sites. When this occurs, an opportunity arises for the remediation of the site at far lower costs than would normally be the case, as the expense of rehandling the large volume of waste material is borne by the remining activity. In addition, remining reduces the need to disturb new areas for the mineral products required by society.

Under Superfund, however, few firms are prepared to acquire old mining sites for remining, since the legal liability to clean up the entire site accompanies such acquisitions.[18] As a result, Superfund perverts the incentive structure in a way that encourages companies to develop new mines in pristine areas free of any past mining wastes or abroad where Superfund legislation does not exist, even where remining the waste piles at old mining sites is less costly.

It is true that mining companies with little or no net worth, that is, firms with questionable financial stability, will be less deterred from

---

[18]Others suggest the same is true for old industrial sites as well. See, for example, Edelstein 1993.

remining since they have much less to lose. One can, however, question the desirability of a policy that discriminates in favor of such firms at the expense of their more successful and financially viable competitors.

It is also true that mining companies will not be deterred from re-mining at those sites where they are already PRPs. Indeed, Superfund liability may actually encourage these firms to remine old sites, even though this is more expensive than mining in pristine areas or abroad since remining bears part of the costs of cleaning up the site. From the point of view of society, this is ideal, because firms in determining where to mine take into account the benefits associated with cleaning up old mining sites as well as the benefits associated with the produc-tion of new mineral products.

Unfortunately, however, Superfund provides such incentives to only a very small subset of mining companies and Superfund sites. At the four sites I am examining, for example, only at the California Gulch site and probably only for two companies—Asarco and Newmont, both of which are PRPs—might one make the case that Superfund provides the proper incentives for remining. At the other sites and for other min-ing companies at the California Gulch site, Superfund not only fails to encourage firms to take account of the benefits from remining for clean-ing up but actually encourages them to exploit higher cost deposits in pristine areas.

The preceding suggests that the liability provisions of Superfund do introduce a number of serious inefficiencies into the way in which the country is cleaning up old mining sites. It is, however, possible that these inefficiencies are partially or totally offset by the incentives Superfund provides for firms to voluntarily undertake the cleaning up of old mining sites not yet listed on the NPL in order to avoid their being listed. Since only sixty of the 100,000 to 400,000 old mining sites in the country are actually on the NPL, this could be a tremendous advantage. In fact, however, very few of these sites are ever likely to be listed on the NPL simply because they do not pose a sufficient public threat due to their nature, size, or location. Moreover, a number of those that might be listed are not prime candidates for voluntary cleanups, because responsible parties with the financial resources required for remediation no longer exist or because the responsible parties are too numerous to agree on an appropriate course of action and the allocation of the costs.

For these or other reasons I am aware of only one mining site—the Bingham Canyon site in Utah owned by the Kennecott Corporation—

where the owner has proposed to EPA a voluntary cleanup to avoid NPL listing. While few details are available to the public, it is clear that despite extensive negotiations the two parties have so far failed to reach an agreement.

This is not surprising. EPA must be concerned about criticism from environmental advocacy groups and others over a bilateral agreement with the company negotiated in private without the participation of other interested parties. Consequently, EPA needs an agreement it can convincingly claim is as good or better from the point of view of the environmental community than could be obtained if the site were listed on the NPL.

Kennecott, on the other hand, has little incentive to agree to such an arrangement, since it might actually do better through litigation. Even if this were not the case, litigation defers the bulk of the cleanup costs for perhaps a decade, thereby reducing the present value of the company's liability by 50 percent or more. Thus, while voluntary agreements from the point of view of society have much to offer, the probability that Superfund liability will encourage the cleanup of many old mining sites under such arrangements does not appear great.[19]

Current and Future Behavior. So far, I have focused on the efficiency and efficacy of Superfund in dealing with past behavior and the mining waste problems that this behavior has bestowed on the present generation. I have argued that past behavior cannot ex post be undone and that all Superfund or any other policy can now do is deal efficiently and effectively with the legacy of past behavior by cleaning up hazardous wastes.

Clearly, however, Superfund can and does affect current and future behavior. Despite the shortcomings I have discussed in terms of cleaning up old mining sites, one might still argue that Superfund internalizes the environmental costs of the mining that is being conducted today, altering the behavior of producers and consumers to ensure that production efficiency, allocative efficiency, and dynamic efficiency are achieved.

This possibility raises two issues. First, Superfund liability, presumably, need not be retroactive to affect the current and future behavior of mining companies. As long as these companies are aware

---

[19]Others suggest Superfund fails to encourage voluntary cleanups at nonmining sites as well. See Edelstein 1993; Connecticut Environmental Roundtable 1993.

that they will henceforth be held accountable for the environmental consequences of their activities, they will have the necessary incentives to treat their mining wastes properly. Probst and Portney, however, suggest that retroactive liability may strengthen the ability of Superfund to affect current and future behavior by "making the strict liability of Superfund a reality." Moreover, they suggest it sends "a clear signal to PRPs that careless management of hazardous substances is not acceptable, even if they are in accord with all the rules and regulations at the time" (1992, 26).

This raises the interesting question of who ultimately should have the responsibility, and the associated liability, for determining the health and environmental consequences of hazardous wastes and for setting the appropriate standards for managing such wastes—the government, producing firms, or a mixture of both? Given the inherent conflict of interest that firms have in setting appropriate standards, a strong case can be made that normally this responsibility should rest with the government. This approach also assures that standards are consistent among firms and avoids the duplication of costs incurred when each firm must acquire the knowledge and expertise necessary to set standards.[20]

Second, it is far from clear that Superfund, which normally requires the creation of a hazardous site sufficiently threatening to public health to merit listing on the NPL before a corrective governmental response occurs, is the most effective approach for dealing with current hazardous waste generating activities. The RCRA, which more directly regulates the current generation of hazardous waste, would appear to be more suitable for this purpose.

Operating Mines and RCRA

Mining and beneficiation generate two billion tons of solid waste a year. This represents nearly 40 percent of the country's total solid waste and is many times greater than the 270 million tons of hazardous substances—

---

[20]Where firms are likely to have superior knowledge of the adverse effects of hazardous wastes and as a consequence become aware of these effects earlier than the government, it may be desirable for the government to share the responsibility and liability for determining proper waste management practices with firms. In the case of mining wastes, however, it is difficult to argue the knowledge of producing firms is superior to that of the government.

that subset of solid waste that poses a threat to human health and the environment—generated outside the mineral sector (Ary 1990).

For these activities, environmental policy can and should ensure that producers and, ultimately, consumers pay the full cost of their goods, including the social costs of pollution. This section first looks at the RCRA and its amendments, which constitute the principal statutes available to EPA and, more generally, the federal government for regulating the current generation of hazardous and solid wastes. This section then considers the performance of federal policy under RCRA in controlling the wastes from currently operating mines.

RCRA Legislation

Congress passed the Solid Waste Disposal Act in 1965 to regulate the treatment and disposal of solid wastes. In the decade that followed, the public became increasingly alarmed about hazardous wastes, and in 1976 Congress approved a set of amendments to the Solid Waste Disposal Act, which along with the original act are known as the Resource Conservation and Recovery Act, or RCRA.

Congress intended that RCRA provide a comprehensive regulatory system for monitoring and controlling hazardous substances from their production to their eventual disposal. RCRA gave EPA the responsibility for identifying hazardous substances, for creating a system of manifests for tracking them from cradle to grave, and for setting performance standards for their treatment, storage, and disposal. Frustrated by EPA's slow progress in achieving these goals, Congress passed the Hazardous and Solid Waste Amendments to RCRA in 1984, setting out in much greater detail exactly what it wanted EPA to do, and when and how it wanted it done.

RCRA and its amendments distinguish between hazardous and nonhazardous substances. Hazardous substances are subject to Subtitle C of the Act, which imposes the monitoring system and other fairly stringent controls described above. Nonhazardous substances fall under Subtitle D of the Act, which leaves the primary responsibility for regulation with the states.

Mining Waste Regulation

From the beginning, EPA had difficulty in classifying mining waste and other large-volume, low-hazard wastes, which were not very suitable to the rather strict and costly regulations imposed by Subtitle C. Responding to the concerns of the mining industry, Congress passed the Bevill amendment to RCRA in 1980, which excluded the solid waste from the mining, milling, and processing of minerals from regulation under Subtitle C until EPA prepared a report on these waste products.

The report, submitted to Congress in December 1985, distinguished between extraction and beneficiation waste on one hand and mineral processing wastes on the other. It concluded that some wastes associated with mineral processing did meet the hazardous criteria for regulation under Subtitle C, but that the very large volumes of waste from extraction and beneficiation did not, for a number of reasons: the sheer magnitude of mining waste compared to the hazardous wastes of other industries; the tendency for mining waste to remain on site; the location of many mine sites away from population centers and in arid regions that reduce the potential for leaching; and the lower risk to human health compared to the hazardous substances of other industries (Holmes 1990; Hoffman and Housman 1990).

Shortly after issuing this report, EPA determined not to regulate extraction and beneficiation wastes under Subtitle C but rather to develop a mining waste program under Subtitle D. EPA recognized that such a program might require federal oversight and enforcement, even though RCRA does not authorize either under Subtitle D.

At about the same time, EPA also proposed to reduce the number of mineral processing wastes exempt from Subtitle C. However, after encountering problems in determining which wastes would remain exempt and which would not, it withdrew its proposal. This prompted the Environmental Defense Fund and the Hazardous Waste Treatment Council to sue in 1986, and two years later the U.S. Court of Appeals determined that EPA's decision to exempt all mineral processing waste was overly broad. The Court ruled that exemption from Subtitle C should be based on the high-volume, low-hazard criteria.

Responding to this decision, EPA issued two final rules during 1989 and 1990 defining which mineral processing wastes met the high-volume, low-hazard criteria. Most mineral processing wastes did not satisfy both criteria, and are now subject to regulation under Subtitle C. The two rules also identified twenty mineral processing wastes (including muds from bauxite refining, residue from roasting and leaching chrome ore, slag and tailings from primary copper processing, wastewater from magnesium processing, and slag from lead and zinc processing), for which a final decision would be made after further study. In 1991 EPA announced that all twenty of these mineral processing wastes would be exempt from Subtitle C. Two—phosphogypsum and process waste-water from phosphoric acid production—would be regulated under the Toxic Substances Control Act, and the rest would be regulated under Subtitle D along with the wastes from extraction and beneficiation.

However, as noted above, there is as yet no regulation of extraction and beneficiation wastes under Subtitle D, in part because RCRA does not give EPA the necessary statutory authority. To encourage public discussion and debate on an appropriate regulatory regime for mining wastes, EPA issued for public comments a proposed approach, referred to as Strawman I, in May 1988, and a revised version, Strawman II, in May 1990. The latter envisages a state-run mining waste program subject to minimum federal standards covering the pollution of surface and ground waters, air, and soils. Performance standards may vary with the site and risks involved, as long as the minimum standards are met. The standards should ensure operations that protect the environment during mining, thereby emphasizing pollution prevention rather than subsequent remediation.

Following Strawman II, and still lacking the necessary statutory authorization to implement a mining waste regulatory program under Subtitle D, EPA in 1991 established the Policy Dialogue Committee. Composed of representatives from industry, the environmental community, and state and federal agencies, the Committee provided a new forum for debating an appropriate regulatory policy for mining wastes.

Conclusions and Implications

Wherever industries pollute there is a need for government policy to ensure that producing firms, and therefore consumers, pay for the full costs, including the environmental costs, of their products. Failure to do so leads to distortions in both production and consumption and, in turn, results in excessive environmental damage.

In the United States, environmental policy under RCRA, even though this legislation was passed in 1976, still does not ensure that mineral producing firms and their customers pay for the pollution they generate. While some mineral processing wastes are regulated, the large-volume wastes from extraction and beneficiation are not.

RCRA, however, is not the only legislation affecting mining and mineral processing firms. At the federal level, there are the Clean Air

Act, the Clean Water Act, federal common law, the Uranium Mill Tailings Radiation Control Act, and, as we have seen, CERCLA. In addition, individual states have their own laws—general mining laws, Superfund laws, tort law, and so on—which regulate and control mining activity. This makes it difficult to assess the consequences of the failure to develop a regulatory regime for mining wastes under RCRA.

The environmental community and EPA contend the failure does have serious consequences, as the present system leaves regulatory gaps, which minimum federal standards would close. State and industry officials, not surprisingly, tend to argue the present system is performing reasonably well.

Since the environmental damage from mining wastes—at least wastes from extraction and beneficiation—remains largely or entirely within the state where it originates, state regulation would at first blush seem most appropriate. The onus of demonstrating that this is not the case would thus appear to lie with those arguing for a stronger federal policy under RCRA.

While the polluter-pays principle can and should be applied to current mining activities, its application to past mining activity is simply not feasible. Many former polluters no longer exist; others do not have the financial resources necessary for remediation. Moreover, the owners and managers of companies that polluted five, ten, or fifty years ago are a different set of people from those who now own and manage these companies. Finally, the ultimate beneficiaries and responsible parties for pollution are the consumers who purchased and used the goods and services whose production generated the pollution.

Despite these considerations, U.S. policy for cleaning up hazardous sites, including old mining sites, under CERCLA attempts to enforce the polluter-pays principle. Unfortunately, the consequences are not benign. While the actual amount spent on litigation and other transaction costs is not known for certain, it probably accounts for between a fifth and a third of the total spent for remediation. Superfund also discourages the remining of old mining sites, often the most effective and least expensive method of cleaning up.

Perhaps of greater significance than the inefficiencies associated with cleaning up old hazardous waste sites is the burden Superfund imposes on current industrial activity. The potential future liability all firms now face as a result of their current activities, even though their activities are in accordance with the laws and the best practices of the day, introduces significant uncertainty and cost. While one might argue this is desirable if it makes firms more environmentally sensitive, it is important to note this benefit is not without cost. Firms ultimately have to pass on to consumers in the form of higher prices the greater uncertainty and higher costs they face as a result of Superfund. There are good reasons to believe it is normally more efficient and more appropriate to make the government rather than firms responsible and accountable for setting environmental standards.

Finally, and more difficult to assess, the claim that Superfund is fair because it forces polluters to pay, when in fact those forced to pay simply have the misfortune of being within the grasp of the law and those who did the polluting are beyond its reach, eventually is likely to foster public disillusionment and cynicism. Indeed, some signs of this are already apparent, less among the public at large than among those more closely associated with the Superfund process. Such disillusionment and cynicism contribute to the erosion of public confidence, not only in current U.S. environmental policy, but more broadly in Congress, the federal government, and the entire political system.

Postscript

The original version of this lecture prompted a number of interesting and thoughtful suggestions, many of which are now reflected in the preceding pages. However, two comments, which are somewhat more difficult to integrate into the main body of the lecture, are examined in this postscript.

The first concerns the costs and problems of now changing Superfund. CERCLA, it is argued, has been in force for over a decade. While a number of companies have come forth and fulfilled their financial obligations under the law, others have resisted and procrastinated. Changing the law so that companies were no longer liable for cleaning up old hazardous sites would either reward the recalcitrant firms or require the government to compensate those firms that have complied. The latter is unlikely, raising a significant equity issue.

In somewhat different terms, the passage of CERCLA in 1980 produced an uncertain but significant increase in the future financial obligations of a number of firms. These expected liabilities led to a reduction in the market value of these firms, adversely affecting stockholders and managers during that period. Changing the liability provisions of Superfund now would have the opposite effect. Uncertainty and expected future liabilities would be reduced for those firms with pending or potential Superfund liabilities, causing a one-time increase in their market values. This would benefit current stockholders and managers, whom many may consider undeserving of such largess.

The point, in short, is that there are often costs in changing public policy, including in many instances the arbitrary confiscation or creation of wealth for certain groups. These costs, which are certainly not unique to Superfund, should be taken into account in assessing the desirability of change. They should not, however, prevent the introduction of new policies or changes in old policies when the benefits to society exceed the costs.

The second comment contends that the lecture, having criticized Superfund, has an obligation to propose an alternative. After all, there is no point in criticizing a policy if there is nothing better.

While there is some validity to this contention, reforming Superfund and the public policies for managing mining wastes will not be an easy task. The issues are many and complex. The analysis here has focused only on allocating the costs of cleanup—which is but one of many relevant considerations—and hence is not sufficiently comprehensive to support a robust proposal for reform.

Yet, before successful reform is possible, a good understanding of the shortcomings of the existing system is necessary. If by examining one important and relevant issue this lecture can stimulate and contribute to the ongoing debate over how best to change current policy, I think it will have served a useful purpose.

In an effort to placate those who may find this response disappointing, it is perhaps worthwhile to conclude by highlighting three policy issues that this analysis suggests are in need of more attention. First, it may be useful for public policy to treat mining waste separately from other hazardous substances, in part because remining may play an important role in the remediation of old mining sites and in part because mining wastes often entail greater volume and lower toxicity than other hazardous substances. Such distinctions have been taken into account in regulating current mining wastes under RCRA, but not in cleaning up old sites under CERCLA.

Second, the liability provisions of CERCLA need careful reconsideration, particularly in the case of old sites where all or most of the damage to the environment occurred before 1980. These provisions discourage remining and in other ways reduce the efficiency of remediation. Moreover, they cannot be justified on equity grounds or the polluter-pays principle since the individuals who did the polluting and who benefited from the pollution cannot now be made to pay.

Finally, there is a need for more knowledge of the incentives public policy can provide for agreements between the government and companies, whereby companies voluntarily clean up old hazardous sites. The number of old mining sites in the United States is very large, and under Superfund only a small number of those sites posing significant threats to human health and the environment are likely to be cleaned up in the foreseeable future. Voluntary agreements, which are much more widely used outside the United States, could help with this problem.

References

Acton, Jan Paul, and others. 1992. *Superfund and Transaction Costs: The Experiences of Insurers and Very Large Industrial Firms.* Santa Monica, Calif.: RAND, The Institute for Civil Justice.

Ary, T. S. 1990. Statement of T. S. Ary, Director, U.S. Bureau of Mines, before the Subcommittee on Mining and Natural Resources, Committee on Interior and Insular Affairs, U.S. House of Representatives, June 19. In *Regulation of Non-Coal Mining Wastes*, published by U.S. House of Representatives. 1992. Washington, D.C.: U.S. Government Printing Office.

Ayres, Robert U. 1992. Toxic Heavy Metals: Materials Cycle Optimization. In *Proceedings of the National Academy of Sciences* USA, vol. 89 (February), 815–20. Washington, D.C.: National Academy of Sciences.

Clay, Don R. 1993. The Superfund Liability and Settlement Structure. Unpublished memorandum, October 14. Don Clay Associates, Inc., Washington, D.C.

Coase, Ronald. 1960. The Problem of Social Cost. *Journal of Law and Economics* 3 (October): 1–44.

Connecticut Environmental Roundtable. 1993. *Superfund Roundtable: Report on Discussion Series.* Hartford, Conn.: Connecticut Environmental Roundtable.

Dower, Roger C. 1990. Hazardous Wastes. In *Public Policies for Environmental Protection*, edited by Paul R. Portney, 151–94. Washington, D.C.: Resources for the Future.

Edelstein, Jan M. 1993. Testimony of Jan M. Edelstein, Director, National Environmental Trust Fund Project, to the National Advisory Committee on Environmental Policy and Technology, August 16. Unpublished testimony.

Hoffman, Steven. 1990. Preliminary List of Extraction, Beneficiation, and Mineral Processing Sites on the NPL, CERCLIS, and Section 304(1) of the Clean Water Act. Internal EPA memorandum to Bob Tonetti, Chief, Special Waste Branch, June 11. In *Regulation of Non-Coal Mining Wastes*, published by U.S. House of Representatives. 1992. Washington, D.C.: U.S. Government Printing Office.

Hoffman, Steven, and Van Housman. 1990. Update on Regulations for Mine Waste Management under RCRA. *Mining Engineering* (November): 1242–44.

Holmes, Christian. 1990. Testimony of Christian Holmes, Principal Deputy Assistant Administrator for Solid Waste and Emergency Response, Environmental Protection Agency, before the Subcommittee on Mining and Natural Resources, Committee on Interior and Insular Affairs, U.S. House of Representatives, June 19. In *Regulation of Non-Coal Mining Wastes*, published by U.S. House of Representatives. 1992. Washington, D.C.: U.S. Government Printing Office.

Portney, Paul R. 1991. Who Should Pay? EPA *Journal* 17 (3):37–38.

Probst, Katherine N., and Paul R. Portney. 1992. *Assigning Liability for Superfund Cleanups: An Analysis of Policy Options*. Washington, D.C.: Resources for the Future.

Western Interstate Energy Board. 1991. *Inactive and Abandoned Noncoal Mines: A Scoping Study*, vol. 1. Prepared for the Western Governors' Association Mine Waste Task Force.

Developing National Policies in Chile

GUSTAVO E. LAGOS

Mining is important to Chile. Despite diversifying its exports during the 1970s and 1980s (mining represented 80 percent of its exports in 1970), Chile still relies on the mining sector for about one-half of its exports. The production of primary copper accounts for some 70 percent of Chile's mining output, which as a whole provides close to 7 percent of its gross national product. Foreign investment in mining has contributed close to one-half of Chile's foreign investment during the last five years and is expected to continue making a fundamental contribution for the remainder of this decade. Approximately one-quarter of Chile's fiscal budget comes from the revenues of the state copper company, Codelco.

During the 1990s, Chile is expected to increase its world share of copper mine production from less than 20 percent in 1990 to more than 30 percent by the year 2000, with an average production growth rate of 7 percent per year (*Boletín Minería y Desarrollo* 1993). At the same time, the rest of the world probably will not increase its output of copper at all. The Chilean mining boom of the 1990s is based on the excellent quality of its ore deposits and the advantageous geography and climate of the north of the country, as well as the general competitiveness of the country's political, social, and economic conditions and the quality of its human resources and infrastructure.

Mining is concentrated in the northern provinces of Chile, a desert zone where there is little agriculture and no forestry. The development of many of the towns and some of the cities was based on mining, and most of the population is still involved in this activity. Yet, mining directly occupies less than 2 percent of the total Chilean labor force of 4.5 million people. Mines, especially large ones, are usually located in remote areas, either at high altitudes in the Andes Mountains or in the middle of the desert. The environmental impacts of mining typically are limited to the local area, with the exception of the atmospheric pollution created by copper smelters, by far the most serious mining impact. Despite mining's importance in the country, its impact on the environment is not as significant as that of, for example, desertification, erosion, urban growth, and industrial pollution.

For the country as a whole, environmental issues are a relatively recent concern. Prior to the mid-1980s, the public and private sectors displayed some environmental concern, but this concern did not lead to action or investment except in a very few cases. Article 19, amendment 8, of the Constitution of 1980, states that every citizen has "the right to live in an environment free of contamination. It is the State's obligation to ensure that this right is not violated and to look after the preservation of nature." Although this declaration was regularly quoted, it had little practical effect. Public (state-owned) companies were among the worst contaminators of the environment, and little was done to change their behavior by the institutions responsible. The National Commission for Ecology was created, but without a budget and with little power it could do nothing but keep a handful of concerned people employed.

Since the mid-1980s, however, public awareness about environmental issues has grown considerably. Across the country, people began to submit legal demands to alleged polluters, many of the demands invoking the relevant but unenforced Article 19, amendment 8, of the Constitution. Legal actions were taken by the following groups:

- the workers of the Chuquicamata mine against their company because of the emission of gases from smelters;
- landlords in the valleys of Puchuncaví and Catemu against the owners of the nearby Ventanas and Chagres smelters;
- the citizens of Chañaral against Codelco, the state copper company, because the tailings from the processing plant of Codelco's El Salvador mine caused a bank of tailings to build up in the bay;
- fruit producers against the Paipote smelter;
- the citizens of El Arrayán against the Disputada mining company because of the danger posed by the expected collapse of the Perez Caldera tailings dam; and

- the olive growers of the Huasco county against Compañía Minera del Pacífico, because emission of particles had caused the olive trees to reduce their production.

In most of these cases the plaintiffs were not successful, although the Disputada mining company agreed to decommission the Perez Caldera tailings dam following an out-of-court settlement. There may have been other out-of-court settlements that were kept secret so as not to encourage other suits. In a few cases, the courts did order that companies install monitors to obtain records of emissions.

During the same period, an increasing awareness of rapidly changing international environmental awareness emphasized the need for the country's most pressing environmental problems to be solved and forced changes in legislation and in policies. In the mining sector, some public companies began to allocate funds to solve the most serious environmental problems. The state-owned companies Codelco and Enami set out environmental guidelines for the first time in 1990, and only then was the environment taken into account when formulating corporate strategy. Although state-owned companies were among the worst contaminators of the environment during the 1980s, companies in this sector did not lag far behind Chile's private industry in demonstrating their regard for the environment.

It is against this backdrop that this lecture examines the progress of environmental public policy toward the mining industry in Chile from 1980 to the present. This lecture analyzes how institutions and legislation evolved in response to the main environmental events of this period. It also assesses how recent events will affect the formulation of future environmental legislation and public policy.

Progress in Environmental Public Policy

The Chañaral Case

The first and most important precedent for Chilean law about environmental matters was set when the citizens of Chañaral won their case against Codelco–Salvador (a division of the state-owned company Codelco) in the late 1980s.

The case involved the Salado River, which carried the tailings of the mineral processing plant of Codelco-Salvador's mine. These tailings totally filled up the bay of Chañaral and killed many sea species. In

1938, the tailings pond of El Salvador, which then belonged to a U.S. company, was filled to capacity, and since then the tailings have spilled into the River Salado. In 1975, the company, which then was owned by Codelco, constructed a canal diverting river water, the tailings included, to Caleta Palitos (a bay), producing the same embankment effect on this bay. Overall, 330 million tonnes of tailings were thrown down the river and then down the canal during this period. During the late 1980s, the Environmental Defense and Development Citizen's Committee of Chañaral took Codelco to court. Soon the court for the district of Copiapó sentenced the company to constructing a tailings dam.

In 1989, the Supreme Court ratified the sentence and Codelco was forced to build a tailings dam, which has now entered operation. This case established an important precedent in Chilean law, showing that despite the inadequate set of environmental laws regarding liquid effluent, companies, even state-owned ones, could eventually be taken to court and forced to consider environmental concerns. This case documents the most serious environmental impact of mining companies on river or sea waters in Chilean history. It reveals clearly that environmental policy was not a priority at all until very recently in Codelco.

President Aylwin's Environmental Program

In 1990, the government of President Patricio Aylwin, elected that year after seventeen years of military rule under General Augusto Pinochet, proposed the following set of environmental actions or goals:

- formulate a national environmental policy;
- face the most critical environmental problems, such as atmospheric contamination in the metropolitan region of Santiago, water pollution (ocean, rivers, and lakes), and massive emissions of contaminants caused by mining and industrial processes;
- set out the basis for an environmental law and dictate the respective legal dispositions;
- promote basic research into environmental matters;
- create an institutional structure to manage the environment efficiently;
- develop an environmental education program; and
- promote an active international policy, which would lead to environmental protection (Concertación por la Democracia 1989).

To carry out this program, the government created the National Environmental Commission (Conama) under Decree 240 of the Ministry of National Territory (June 1990). The aim of this interministerial commission is to look after the "study, proposal, analysis, and evaluation of all matters related to environmental protection and conservation."

In spite of this policy for environmental action, it is evident that progress since 1990 has been mixed. Partly this is because the environment has not been one of the government's main priorities. Conama is located institutionally within the least important ministry and so did not carry much weight until 1992 when, encouraged by the possibility of a free trade agreement with the United States, the most influential minister in the government, the finance minister, made his first speech about the environment. It was not until September 1992 that the government sent to Congress the draft of the Environmental Framework Law, known by its initials, EFL.

Environmental Legislation and Institutions

A significant number of laws exist at present that deal with environmental issues (Lagos, Noder, and Solari 1991). Conama has cited 2,200 such laws and regulations (El Mercurio 1992b), some of which date back to the beginning of the century. The laws as they stand require neither an environmental impact study nor a base-line study for new projects. The laws also do not contemplate land reclamation or the abandonment of sites—whether industrial, mining, or of some other type. There are no standards regulating liquid and solid effluents or soil quality. Legislation concerning dams built up from the tailings process is not up to date, and new legislation prepared by Sernageomin (National Geological and Mining Service) is being discussed at present. The water quality standards date from 1970 and 1978 (Supreme Decrees 357 and 1333 respectively) and are fundamentally the same as those applied by the U.S. Environmental Protection Agency (EPA).

The most important advance in legislation related to mining under the Aylwin government was Supreme Decree 185 of January 1992, which regulates emissions of sulphur dioxide, arsenic, and particles from fixed sources outside the Santiago region. This decree replaced Resolution 1215 of 1978, which had had little practical effect. Decree 185 will allow Chile to reduce its sulphur emissions "threefold" by the end of the decade, by which time every smelter in Chile should be complying with the standards, provided that the financing for it is provided by the state or that the property of the companies is transferred to the private sector (Solari and Lagos 1991; *Boletín Minería y Desarrollo* 1992).

The next sections deal with the way in which the institutional system has tackled environmental problems in the past and the progress made during President Aylwin's government, especially regarding the philosophy of the recently approved legislation. Then the legislation that should be passed in the 1990s will be considered, as well as the challenges the legislation will address.

Overlap of Responsibilities

There are also many institutions responsible for the environment with duties and rights that overlap. For instance, the control of the quality of the country's water resources is carried out by seven different institutions: the General Direction of Water (DGA), the Agricultural and Cattle Service, the National Forestry Commission, the Directorate of Maritime Territory, the Sanitary Water Authority, Sernageomin, and the Ministry of Health. These institutions belong to different ministries and therefore often have different sets of criteria for dealing with matters related to the environment. As a result, on many occasions there have been conflicts, which not only damage the possibility of protecting the environment but also cause delays in issuing the permits necessary for companies to start operations. More than once, an application for a permit to build, for example, a tailings dam took several years to be reviewed.

Weak Procedures

Additionally, the procedures and methods employed by the various government agencies have been hampered by a lack of funds, insufficient and poorly trained personnel, excessive centralization, inadequate standards, and a decision-making process in which decisions often are left to the judgment of a single person in one of the regions or in the central headquarters.

In 1990 there were ninety-eight employees looking at environmental matters in the four most important agencies responsible for the control of water quality, agriculture, and forests in Chile, and only ten employees dedicated all their time to environmental matters (Lagos, Noder, and Solari 1991).

Until recently the DGA, the main agency controlling water quality throughout the country, did not have records of mercury content in rivers, even though close to 300 very small scale gold mines are authorized to operate and many of them use amalgamation as the process to obtain gold from ores. This lack of records occurred due to lack of funds (Muñoz and Lagos 1990).

Another example of weak procedures is that, for several decades prior to 1990, none of the information that DGA possessed about water quality had ever been analyzed in a systematic way, so decisions to close plants were usually taken (and they were taken occasionally) after observing the numbers without reference to a model of any sort. For example, the closure of El Indio's mineral-processing plant was due to cyanide contamination in its tailings dam (and thus in Río Malo) and was carried out under articles in the sanitary code. The Río Malo closure was justified, since the levels of cyanide had been shown to increase dramatically downstream of El Indio's treatment plant at the beginning of the 1980s. Eventually, this company had to modify its process in order to continue gold production. It is clear, however, that in many instances, treatment plants, mines, tailings dams, smelters, and leaching operations were not closed when all the evidence indicated that they should have been closed.

Yet another example of the precariousness of administrative procedures is the case of the existing water quality standard, which is based on soluble contents, as are the standards of the EPA. But the measurements of the DGA are of insoluble content. Thus, in practice, it is impossible to establish a parallel between the records and the standard.

It should be said that some countries base their standards on insoluble contents, which in the long term seems a more secure way of protecting the environment. However, an insoluble content standard should be based on a complete knowledge of the geochemistry of a certain region. If the soil is acidic, such as that around many Canadian lakes, then it is advisable to use an insoluble standard, since most of the metal contained in the water could be dissolved and absorbed by plants, animals, and so on. However, if the soil is basic or neutral, the use of a soluble standard could be acceptable in rivers and lakes, since much of the metallic content is mechanically transported through the water and most if not all of the metallic content should precipitate at some stage.

In some Chilean cases, such as the Malo River, there is acidic soil and the base content of metals is several times higher than the standards (Lagos 1990). However, approximately twenty kilometers downstream, the soil becomes basic, so all the metal precipitates and the environmental impact is null. Some of the rivers in northern Chile have a base level of metallic elements that is higher than the existing standard (Decree 1333); in practice, then, if the DGA were to force all the mining companies to observe the decreed standards, many could not continue to operate and the new mines would find it difficult to obtain permits.

Similar situations exist at the agencies regulating the air, soil, and other matters related to the environment.

Agencies As Good As the System

The intervention of Chile's various agencies in the regulation and control of the environment has been no better than the institutional and legislative system that supports them. Most of the environmental incidents or events that occurred during the 1980s came to public notice because they were simply too visible or because the local communities felt threatened and they called the press. Likewise, almost all the environmental episodes registered during the 1980s were brought to the public arena by local communities or by environmental organizations, and not by the state regulating agencies. It is not that these agencies could not perform control actions or that they did not order in many instances the closure of processes or even plants. The Chilean environmental legislation, although scattered, could have been used in many cases by the state. The monitoring was available—especially regarding water quality—but the analysis of the data, the decision-making procedures, and in the end the political will were not there. The laws and regulations in most cases established rules, without specifying deadlines for compliance or fines for lack of compliance. For instance, the law that regulates liquid effluents dates back to 1916 and says that the industries cannot dispose of liquids that carry any contaminants.

Advances since 1990

Nevertheless, important advances have occurred under the present government, especially in the mining sector, mainly because the government has not waited for the institutional and legal changes that are necessary in many areas in order to reach environmental solutions. Instead the government has worked with the concept of ad hoc commissions, established by goodwill between the different ministries and other institutions involved, and also with the open cooperation of companies, private and public. Companies, contrary to what may be understood, would like legislation and clear rules soon. Most of the procedures applied to new projects today regarding environmental impact studies, base-line studies, liquid and solid effluents, even abandonment procedures, have been devised without reference to specific legislation, because the legislation simply does not exist or is not coherent enough. Although this practice should be very useful in establishing realistic legislation in the future, it is also based on a concept much feared by the country's legal system, which is that of leaving ample room for the discretion of the opposing parties, particularly the discretion of state officials, who will no doubt change with future governments.

A second advance has been the increasing use by state regulators of some of the 2,200 laws and decrees already on the books related to the environment. This is another track open for the discretion of state officials. The legislation is so abundant in some areas that no doubt one could find within it almost anything required to impose environmental control in a specific case. There is also no doubt that if a company sought to respond to one of these administrative measures with a court suit, the likelihood of finding another decree or law that annulled the first one would be high. And so on.

Until 1992, Chilean environmental legislation was based on command-and-control criteria. However, Decree 4 of the Ministry of Health, from March 2, 1992, establishes a maximum air emission standard for industry in Santiago's metropolitan region; this standard is much lower than what is emitted at present and should be achieved gradually in order to meet all of the decree goals by December 1997 (CEPAL/PNUMA 1992; Comisión de Medio Ambiente 1992). It also establishes a methodology for calculating emission rights, which can be attributed to each company located in the region and which can be sold or bought.

The Environmental Framework Law

Three new concepts in Chilean law were to be discussed by Congress during 1993, after President Aylwin sent the EFL project to the Senate on September 14, 1992 (El Mercurio 1992c; Aylwin 1992).[1] This should

---

[1]This lecture was presented in late 1992. In January 1994, the EFL was approved by the Chilean Congress.

also be the first law that will regulate the environment as a whole, serving as a reference to the 2,200 regulations that exist at present.

The first concept is responsibility for environmental damage, which was not contemplated explicitly in previous Chilean legislation. The EFL project establishes the requirements for environmental reparation when environmental damage has been produced. Under EFL, when a suit is filed against a violator of environmental regulations, the plaintiffs will be able to obtain indemnification if they can prove that damage to their property or quality of life has occurred, following the precept established in the Chilean Constitution that all citizens have the right to live in an environment free of pollution. The definition of quality of life that applies in a particular case will presumably have to follow the same intricate pattern that it has followed in developed countries. Some major universities in Chile are beginning research on the evaluation of environmental damage. Thus the strategy of companies when suits are filed will be quite different from strategies observed in the past; for instance, in the case of smelters, court orders used to go as far as requiring the companies to monitor the environment, but they seldom ordered the closure of plants and never ordered indemnification.

It is relevant to discuss briefly the case of the Compañía Minera del Pacífico (CMP) at Huasco, since the discussion of the case in the court may coincide with the approval of the new EFL legislation. In August 1992, CMP, which is located in northern Chile, was found guilty by the Supreme Court of contaminating the Chapaco Bay with its tailings and the Huasco olive tree plantations with particles emitted from its iron pellet plant. This company had been the subject of numerous suits by different groups during the last fifteen years (El Mercurio 1992a; La Epoca 1992). According to the plaintiffs, the pellet plant has emitted particulate material that has covered the leaves of olive trees in nearby plantations, blocking photosynthesis and bringing down olive production to negligible levels. Parallel to this, the plant has been disposing its tailings directly into the sea at Chapaco Bay, thus producing a considerable negative impact on the bay's flora and fauna. The plaintiffs—olive tree farmers and the fishermen and divers of Chapaco Bay—have claimed damages of US$64 million (El Mercurio 1993). The Supreme Court ruling ordered the company to reduce its levels of particle emissions in order to comply with Decree 4 of the Ministry of Health and to eliminate disposal of contaminated effluent into the sea, all these within one year of the sentence. But the sentence established no compensation for the plaintiffs. EFL legislation should contain criteria that will change the results of such court actions in the future, and perhaps it will have an effect on this case as well.[2]

The second concept in EFL is a requirement for "large" new projects to have an environmental impact assessment study (EIS) prior to obtaining a permit. As mentioned previously, this is already a working practice in the mining and other economic sectors in Chile. However, the establishment of this requirement by law will be quite complex, due to the multisector way in which environmental decisions are adopted at present, involving many different state agencies and ministries. This is an aspect in EFL where the competence of Conama would intersect the competence of ministries. Conama probably will retain its coordinating role, with direct access to the president but it will not take into its hands the decisions that at present are the responsibility of the ministries. Moreover, the practice of the last two years has been to assign increasing power to the regional authorities. It is likely that an EIS will be revised and decided once and only once in multilateral commissions organized at the regional level.

A third new concept introduced by EFL is what is known elsewhere as the polluter-pays principle (see John Tilton's lecture in this book). However, EFL is quite vague about this point, and it refers to future legislation on the subject, especially as related to tradable emission rights, which are already contemplated in Decrees 185 and 4 (for the Santiago region). In Santiago, tradable emission rights apply to all the metropolitan area, which constitutes a "bubble" with total particle emissions defined by Decree 4. Some companies are already setting the basis to trade their rights. The new clean air act adopted in 1990 in the United States also establishes the concept of tradable emission rights, and these have been made effective in relation to sulphur dioxide emissions and acid rain.

EFL maintains the basic structure of the state, with the regulating agencies staying as they are today, within different ministries. In practice, however, EIS approval decisions are being shifted to the regional level, where a single-window state procedure involving companies and individuals should operate in the future. But it is not clear that this type of regional organization would be applied to all sorts of environmental permitting and other aspects of environmental control. In the

---

[2]The EFL legislation that was passed does indeed contain such criteria.

short term, it is likely that it will not, considering the complexities of such actions.

As to the future direction of the institutional system, there are two broad options for environmental institutions, each supported by sectors within the country that hold opposing views regarding the balance between growth and environmental protection. While the current trend has been mainly backed by more liberal sectors, whose concerns are as much with growth as they are with protection, the opposing view, which would create a centralized environmental agency or ministry, has been supported by groups who are more cautious about growth and its environmental impacts. (It should be added that the sectors responsible for implementing environmental actions at a government level have been the more liberal ones.) The discussion between these positions extends to the entire range of environmental problems.

New Legislation for the 1990s

As should be clear from the discussion above, several environmental issues have not been covered adequately by current legislation. In most cases, it is thought that legislation will be created at some stage during the coming decade. The following sections consider areas of environmental concern and prospective legislation regarding those areas.

Tailings Dams

Sernageomin, under the direction of the Ministry of Mines, has proposed to the government a project that modifies the present legislation governing tailings dams. The changes propose to do the following:
- replace Supreme Decree 86 of July 31, 1970;
- create a single-window procedure through the creation of an interministerial commission;
- improve procedures to construct and obtain a permit to operate a tailings dam;
- create new procedures for abandoning a tailings dam, which would also apply in emergencies;
- leave the final authorization and supervision on permits for tailings dams to the director of Sernageomin;
- include a very general methodology to evaluate the impact of solid

and liquid effluents and clarify the procedures to obtain a permit to dispose of these effluents.

However, the Sernageomin legislation could not possibly define standards related to liquid and solid effluents. The clauses concerning information on liquid and solid effluents and waste indicate that the State wishes to know how these effluents are being treated or disposed of, possibly with the intention of making more specific rules in the future. Especially important should be reporting the transport of material, whether it be sterile, ore to be leached, leaching wastes, tailings, slags, dust cakes, or other solids. At present, no legislation requires mines or plants to report such transport, except in the case of tailings.

It is desirable that tailings dam legislation should consider land reclamation where this is applicable in relation to the abandonment of tailings dams, leaching dumps, sterile dumps, leached wastes, mines, slags and dust cake dumps, and other type of solid and liquid dumps.

Water Quality

Legislation on water standards should be made more realistic than it is currently, and it should take into account the natural composition or base levels of rivers and streams and dictate special decrees for such cases. For instance, it would be absurd if Decree 1333 were applied to Río Loa upstream of Calama and Chuquicamata, since Codelco has a water treatment plant that extracts arsenic from the natural water of the river to make it drinkable for the population of Chuquicamata and Calama.

Legislation will possibly continue to be based on soluble contents rather than on insoluble ones, which the DGA advocates, at least until there is more knowledge about soils and geochemistry in Chile. It seems important as well that compatible procedures be established for monitoring water by the different agencies that have jurisdiction over this matter.

Soil Quality

Soil standards should also be defined, though legislation in this area could take at least five years due to the lack of data regarding soil quality. Areas such as Catemu (the community where the Chagres smelter is located), Chuquicamata, and Calama, have plenty of mining company data about soil quality, but the data have not been published. Areas like the valley of Puchuncaví, where the Enami Ventanas smelter is located, have very little published data, making it impossible to draw definite conclusions about trends in the quality of the soil (Gonzalez and Bergqvist, 1985a, 1985b; Chiang and others 1985; Gonzalez 1992). An even more uncertain situation regarding the monitoring of soil quality occurs close to other smelters like Potrerillos, Paipote, and Caletones.

Regarding the impact of liquid mine effluents on soil quality, the existing studies are not only scarce but also poor. A number of studies have analyzed the quality of water in different river basins, but only three have attempted to link water quality with soil quality (Muñoz and Lagos 1990; Luna and Lagos 1990, 1992). This is due to the lack of organized and systematic data regarding soil quality. Geochemistry information is also scarce and heterogeneous (and not public). Until such information is available, legislation about soil quality seems very unlikely.

Abandonment

Legislation about abandonment is likely to be created before the end of this decade, since several abandonment cases will arise during this period. There were few reported environmental events during the last decades regarding abandonment and land reclamation. Flooding seems not to have been an issue in Chilean mining.

Several large nitrate mines in the Atacama desert, the driest ecosystem in the world, were abandoned during the first half of the century, and yet, fifty years later, there is no evidence that such action carried any environmentally detrimental effect, except the disruption of the landscape. But even from this point of view, the abandoned mining towns of the nitrate mines constitute today a place that many visit, since the mining towns are like industrial museums of the beginning of the century.

Not all the mines in northern Chile are located in the desert itself. Many are in the Andes mountains, close to the salt lakes or on the altiplano ecosystem, a plateau that has unique flora and fauna. These mines could constitute in the future a source of disruption to the environment, especially when abandonment occurs.

An example of mine abandonment occurred recently with the termination of the Choquelimpie gold mine, which is owned by Shell, Citibank, and Northgate and is located in a national park just above the altiplano, at an altitude of 4,200–4,800 meters. This mine used cyanide leaching and was run according to the environmental rules of Shell, which are indeed much more strict than Chilean legislation. It could be that the procedures to close this mine will become a model for future Chilean legislation, especially on the issues of who pays for cleanup and who is responsible after closure or after re-opening the mine and plant under a different owner.

The mines located from Chile's central region toward the south constitute an altogether different case, since the rainfall in these regions increases toward the south, reaching levels of more than two thousand millimeters per year in the austral region. The large- and medium-sized mines located here include Codelco-Andina (sixty miles northeast of Santiago), the two mines of Disputada (north and east of Santiago), Pudahuel (west of Santiago), Codelco-El Teniente (sixty miles southeast of Santiago), and the coal mines located in Lota and Coronel. In the extreme south, close to Punta Arenas, is the lead-zinc mine El Toqui, which is located in the unique ecosystem of continental Aysen with more than two thousand millimeters of rain per year and with abundant virgin rain forest. The abandonment of any of these mines should be looked at with extreme care since events of consider-able proportions could be expected, such as flooding and acid and metal contamination of river basins and the ocean.

Very soon, the coal mines located in Lota and Coronel may well be abandoned, because they have uneconomical production costs and have been running at a deficit during the last few years. The even-tual closure of these mines should bring about the need to reconfig-ure the region industrially, because coal mining has constituted the only economic activity for tens of thousands of people since the last century.

The most important case concerning closure of a mining installation occurred during the 1980s. The Perez Caldera tailings dam stored the tailings from the Los Bronces mine, which was owned by the Compañía Minera Disputada de Las Condes, an Exxon Minerals mine since 1978. In 1936, Chilean authorities approved the construction of Perez Caldera tailings dam number one, adjoining the Los Bronces mine, located at 2,800 meters above sea level and just thirty-two kilometers from Santiago. The San Francisco River bed was chosen to deposit tailings, and a tunnel, destined to alter the course of its waters, was con-structed, leading the water parallel to the dam and returning it to the river, downstream of the dam.

In 1978, the construction of Perez Caldera tailings dam number two was approved, along with the construction of a second bypass tunnel. In 1987, construction of a third tunnel with a larger capacity, to act as an overflow for the second tunnel in case of an emergency, was started. However, before work on number two had finished, in December 1987, the second tunnel was blocked by a collapse caused by thawing. This impeded the evacuation of water, which ended up in the dam instead of the canal. During the five-day emergency, more than 300,000 cubic meters of water were collected in the dam, which meant that the water level was less than one-half meter from the top edge.

The El Arrayán Town Council, a Santiago neighborhood located in the river basin in the high part of the city, presented a protection appeal to the Court of Justice, encouraged by an alarming report from a firm specializing in tailings dams. The report pointed out the danger of eventual collapse as a result of floods and/or earth tremors, so that a great mass of tailings could flow and reach the high part of Santiago. Likewise, the residents of El Arrayán were concerned about the effect of these tunnels on the downstream water level, since changes in water level could damage the local flora and fauna.

The Disputada Company and the El Arrayán residents reached an agreement. The company agreed to evacuate the tailings dam when the expansion project of the Los Bronces mine materialized. The evacuation operation would be carried out by transporting repulped tailings through a pipeline to the Colina locality in a dry-region plain, where the new tailings dam for the Los Bronces expansion would be built. The removal of seventy-six million tons of tailings contained in both dams would take twenty years and cost close to US$100 million, including operational and investment costs.

The residents promised to withdraw all lawsuits and accepted the use of the El Arrayán Brook whenever the rise of the San Francisco River made it necessary, limiting the discharge to thirty-five cubic meters per second of water and restricting access to the river to the thawing season (five months a year). In a similar way, the agreement established that once the dams were emptied, the course of the San Francisco River would be returned to its natural river bed and the third tunnel would be dismantled.

There was no precedent in Chilean legislation regarding this type of situation. Nevertheless, the Disputada Company acted swiftly to reach this agreement out of court, thus showing its concern for environmental problems. But what would have happened if the Los Bronces mine expansion had not gone ahead? Who should have paid for the evacuation of these dams? What would have happened if the Disputada Company had not reached an early agreement with the citizens of El Arrayán? Chilean legislation is not prepared to face these types of problems even today, and it seems imperative that such legislation should be created soon.

Water Use

It seems necessary to revise the water use legislation to employ this resource more efficiently in the desert areas in northern Chile and also to preserve unique ecosystems like the ones that exist in the altiplano. The increasing demand by mines, mining plants, industry in general, and growing cities for more water has already made this resource scarce. At present, there is the perception that water has not been administered with an eye toward the long-term development needs of northern Chile. Nevertheless, it should be pointed out that the DGA has diligently looked after the use of water for many decades and that sophisticated models for the use of water in many river basins have been developed.

What DGA may not have taken into account is that, in many cases, technological options are available that allow water to be used more efficiently, though at a higher cost for mining companies. This seems one of the aspects that new legislation should consider. Another aspect is related to the allocation of water permits or rights, which in some parts of the country seem to be concentrated in very few hands, usually those of companies that have had mining or energy-producing activities for many decades.

Atmospheric Arsenic

The standard for arsenic in the atmosphere has not yet been defined, even though the deadline for establishing such a standard was July 1992. This is due to a difference of opinions between the Ministry of Health and the Ministry of Mining. A study made for the Ministry of Mining showed that an upper limit for the annual average of the standard could be 0.1 µg/m^3. A daily basis would be the same as the annual average. However, the Ministry of Health advocates a standard of 0.02 µg/m^3. (The EPA does not have an arsenic standard for the air, but at least one U.S. state, Washington, has an arsenic standard of 0.2 µg/m^3.)

The average concentrations measured in locations next to emission sources like Puchuncaví, Calama, Chuquicamata, and Tierra Amarilla are above these levels (Asociación Chilena de Seguridad 1991). It is apparent from the analysis of the arsenic emission data that considerable investment (additional to those of acid plants) would be required for all the state-owned smelters if the standard is fixed at 0.1 µg/m^3. One of the options being sought at present is to define such a standard and, simultaneously, realistic deadlines so that all the smelters can gradually approach the standard. However, if the standard is fixed closer to the value advocated by the Ministry of Health, the base levels in some localities would be above it and some of the smelters would eventually have to be closed, since compliance would be outside the currently known technological solutions.

Discussion

Local and International Standards

Mining companies obtained permits in much shorter times in 1993 than they did five years earlier; they also obtained their permits in less time than their counterparts in industrialized countries. Yet the environmental impact studies of these mining projects seem to show that the environmental requirements are no less demanding than those employed in industrialized countries.

The reasons for such results would seem to be the geography and climate of northern Chile, which are more amenable to mining and metallurgical operations than possibly any other place in the world. But is this really true or is it only a justification? Are the environmental standards in Chile as demanding as those used by industrialized countries? This question applies not only to air, water, and other "hard" standards, but also to subjective standards that the public and government perceive should be applied. (These subjective standards will be referred to henceforth as *environmental gauges*.) Is it true that transnational mining companies come to Chile because they can apply here lower environmental standards than in industrialized countries? Let us try to respond to these questions.

When one makes the most basic calculations as to how much companies in industrialized countries spend on the extra time needed to obtain permits, it is obvious that it is advantageous for them to come to

Chile. But this could well mean that the system is more efficient, not necessarily that it is less demanding.

One of the arguments frequently cited is that Chile's population needs mining development, especially at a regional level, so that some environmental impacts may be willingly overlooked. In Chile, as in many other countries, the lower income strata of the population—those who are defined as poor[3]—are less concerned with environmental matters than with basic social needs, such as housing, health, education, and the availability of jobs. Because this segment of the population makes up nearly one-third of the total inhabitants of Chile, its specific weight in national policies is very relevant.

In economic terms, the perception of Chile's poor is that the social benefits of reducing pollution are close to null when the social costs of such a reduction mean that new jobs will not be created, salaries will not be increased, social infrastructure will not be built, and so on. The classic example is the smelter where the trade union sued the company for not controlling pollution, so the company announced that the immediate solution was to close the smelter and sell the concentrate rather than processed metal. An agreement was reached very soon afterwards. The environmental gauge employed by the public can be lowered when social benefits are at stake.

Thus one might conclude that despite the absence of a deliberate policy by the companies or by the government, the environmental gauge employed by the public is lower in Chile. But this does not mean that the environmental impacts of new mining operations are greater than the impacts accepted by industrialized countries. Indeed, as has been discussed, geography, climate, and other factors are very favorable for mining in northern Chile, and the country may reach the next century with very little need to rehabilitate the land around mining operations that are at very high altitudes and/or in remote desert locations, or to be very much concerned about the eventual abandonment of those same operations, since their potential impact would be, according to present scientific data, very small.

Regarding the existing "hard" standards for air and water, it has already been mentioned that Chile's standards are essentially the same as those employed by the U.S. EPA. But neither the population nor the government of Chile has perceived yet the strong need for the

---

[3]The Ministry of National Planning defines "poor" as those families that earn less than a certain amount per month.

creation of most of the missing legislation and standards. When this new legislation comes into effect, mining companies will face a level of difficulty in their projects that comes closer to the difficulties that they face today in industrialized countries.

It sometimes is argued that the large mining investment occurring in South America—and especially in Chile—is due to greater leniency toward environmental standards and regulations than national laws actually permit. This argument suggests either that large multinational companies are acting in collusion with local governments and authorities or that local authorities are ignorant about what mining companies are doing. If these assumptions were true, however, shareholders of multinational companies would have to be informed of this deliberate policy at some stage, and it is difficult to see how such actions could be accepted. Moreover, during the past ten years, the government of Chile, and for that matter the public as well, have consistently shown themselves to be more strict with foreign companies than with Chilean-owned companies on environmental matters.

Efficacy of Air Quality Legislation (Decree 185)

Atmospheric smelter emissions are the main environmental problem related to Chile's mining. Seven copper smelters, five of which are state-owned, are the source of these emissions. Only the two private smelters comply with Decree 185, which regulates air quality, while the five state-owned smelters must still make substantial investment in order to comply.[4]

Why have the state-owned companies not been able to comply with this regulation? It is obviously a question of priorities. The government and the public seem to consider that the social benefit of making the private companies comply is high in relation to the social cost, while they seem not to perceive this as the case for state-owned companies. But, it should be added, no one has measured or calculated either the social benefit or the social cost in any of these cases.

The position advocated by the Aylwin government is that the social benefits of appropriating a maximum of the profits from these state-owned companies for direct social projects is greater than reinvesting

---

[4]The total sum required by Codelco and Enami to comply with Decree 185 was estimated to be US$1 billion in 1991. This sum does not include arsenic air emission compliance by the smelters.

those profits in environmental projects.[5] However, this is not the only or the most important reason. A significant motivation for not allocating the required funds is that the economic authority doubts the efficacy with which the state-owned companies are managed within their institutional restrictions. Moreover, there are strong indications that the economic authority does not believe these companies should remain state-owned for long. That this matter has not been voiced is consistent with the important political weight that the more conservative sectors of the country have.

Approaching the mid-1990s, it seems illogical that a country like Chile, which modernized its economy earlier than any other Latin American country, is still troubled with the privatization of its copper companies. Why should the state keep for itself a productive role in this sector and only in this sector? The answer should be found not in economic or political theory, but rather in the popular sentiment that to some extent still identifies the property of Chile's copper mines with its independence and its freedom. In this sense, copper in Chile at the beginning of the 1990s is like Britain's coal at the end of the 1970s.

The conclusion, then, is that the efficacy of Decree 185 is dependent on deep political factors, which will possibly untangle themselves during the present decade but which will certainly retard achieving full compliance with this decree.

The Institutional Options

EFL's institutional approach, which consists fundamentally of retaining the status quo, has many shortcomings regarding the protection of the environment. However, the main alternative to EFL, which is to concentrate all or most environmental powers on one institution, is much resisted. Little discussion has taken place during the Aylwin government regarding the reorganization of the state, since its policy priorities were related to the political, economic, and social stability of the transition to democracy. If any significant change comes in the reorganization of the state during the next government (1994–2000), most likely it will be motivated by the need to make Chile into a more competitive economy, more open to world trade and investment. Change will be motivated by the need to join the North American Free Trade Agreement

---

[5]The only exception to this rule in the period 1990–1993 was the investment made in the Chuquicamata smelter, which was selected because of its larger environmental impacts.

(NAFTA), MERCOSUR (the free trade agreement that includes, so far, Brazil, Argentina, Paraguay, and Uruguay), and the Pacific Economic Cooperation Agreement. Change will also be motivated by the need to keep up with the same levels of economic growth achieved in the period 1983–93, which averaged more than 6 percent per year.

Follow-up legislation to EFL should include three additional features if environmental controls are to work with efficacy: decentralizing power, coordinating all agencies that have environmental responsibilities, and empowering the community.

The decisions related to the study and approval of environmental impact studies have been shifted from the central to the regional level. This has signified much involvement by central authorities and by consulting firms at regional levels in the study of each EIS. It could not be otherwise, since regional officials generally have not acquired yet the training or the expertise to judge whether a particular EIS has the appropriate reference terms or whether its requirements are suitable. It should be added that reference terms are at present quite discretionary, especially in the areas with no precedents or environmental legislation.

The idea of shifting decisions to the regional level would seem appropriate if it involved, in the medium term, all the areas concerned with the environment—not only EIS decisions—and the creation of effective coordination between various agencies at this level. Hence, this could become a true single-window and coherent environmental feature of the institutional system.

Another issue emerging at the regional level is how the community participates. Although there are no established channels for community participation in environmental matters, in recent years public participation has influenced some EIS decisions concerning foreign mining companies. Yet the public has had nothing or little to say in any of the EISs made in the same period by Chilean private and public mining companies. This should not be taken as an indication of the quality of the impact statements, but rather as a political bias introduced in the environmental arena. One has but to go back to the mid-1980s, when a special environmental decree was imposed on one of the six copper smelters, the single one that belonged to a foreign company, Exxon Minerals. Until 1992, this smelter was the only one that complied with environmental regulations.

Thus foreign companies may have a tough time in the future. What appears to be of great relevance for the fairness and credibility of the

Chilean system is that a formal procedure for the participation of the public is introduced in the discussion of environmental impact statements.

Concluding Remarks

There is a striking coherence in what has been happening with attitudes and policies toward the environment in Chile since 1990. Despite the increased importance of environmental matters in national policies during this period, these are still far from constituting priority. The motivations that will likely drive the state to improve its environmental laws and institutions—and as a result to solve its most serious environmental problems—have more to do with economic competitiveness than with specific concerns about the environment. Either way, this could indicate a very positive trend, since international experience seems to show that open and competitive economies are friendlier to the environment than closed and noncompetitive economies.

The dissension regarding institutional reorganization is only an expression of a deeper difference concerning the environment between those who put an emphasis on growth as much as on environmental protection and those who are more cautious about the implications of economic growth. It is likely, but not yet demonstrated, that the latter sector is related to more traditional economic theories. The complexity lies in that both factions cut horizontally across political parties, center and left. During the discussion of EFL, concepts such as zero emissions, contamination as "any change in the state of the environment," guilty-if-not-proven-innocent, and so on have been analyzed and may eventually be introduced in legislation.[6] However, it is early yet to predict whether EFL and its subsequent legislation will be coherent enough to be applicable.

Nevertheless, Chile has the opportunity to become a leader in the environmental arena. The new legislation can be designed realistically, and its application can be achieved gradually, if only the consensus politics, which have been applied so successfully during the transition of the country to a democratic system, continue to be applied to all the modernizations that the country requires, including environmental

---

[6]As it turned out, these concepts were not included in the legislation as approved in 1994.

matters. Another possible favorable omen for success in this endeavor is that Chile has been able to start almost from scratch on matters concerning the environment. This may seem contradictory, since at present there are more than 2,200 laws and regulations related to the environment. However, most of these laws have never been applied, and some of them have been applied only too recently, due to an emerging political will. The point is that environmental laws and regulations, although they may be written, have only been used when there is a political will to do so, and, moreover, these laws and regulations are more likely to be used and used properly if they are reached by a process of consensus. There is no need to follow in the footsteps of countries that have adopted inapplicable environmental legislation or whose environmental legislation has involved enormous social and economic costs.

References

Asociación Chilena de Seguridad. 1991. Anales de la Primera Jornada sobre Arsenicismo laboral y ambiental, II Región. August.

Aylwin, Patricio. 1992. Speech to Congress regarding the Environmental Frame Legislation Project. September 14.

Boletín Minería y Desarrollo. 1992. Published by the Center for Copper and Mining Studies (CESCO), Santiago, Chile. July.

―――. 1993. Published by the Center for Copper and Mining Studies (CESCO), Santiago, Chile. March.

CEPAL/PNUMA. 1992. Instrumentos económicos para la política ambiental: documentos seleccionados. LC/R.1138. Santiago, Chile: CEPAL.

Chiang, J., P. Cornejo, J. Lopez, S. Romano, J. Pascual, and M. Cea. 1985. Determinación de cadmio, cobre, manganeso, plomo, hierro, cinc, y arsénico en sedimento atmosférico en la zona de Quintero, V región, Valparaíso. Bol. Soc. Chil. de Química 30 (3): 139–158.

Comisión de Medio Ambiente. 1992. El uso de permisos de emisión transables en el control de la contaminación atmosférica. Report no. 187. Santiago, Chile: Centro de Estudios Públicos.

Concertación por la Democracia. 1989. Electoral Program. Santiago, Chile.

El Mercurio. 1992a. Article, August 9.

————. 1992b. Article, August 23.

————. 1992c. Article, September 11.

————. 1993. Article, March 20.

Gonzalez, S. 1992. *Riesgo Ambiental para los suelos de Chile*. Report La Platina Number 70. Santiago, Chile: Instituto de Investigación Agropecuaria.

Gonzalez, S., and E. Bergqvist. 1985a. *Evidencias de contaminación con metales pesados en Puchuncaví*. Fourth Symposium on Pollution, National Institute for Agricultural and Cattlery Research. Santiago, Chile: Instituto de Investigación Agropecuaria.

————. 1985b. *Suelos contaminados con metales pesados, efectos sobre el desarrollo vegetal*. Fourth Symposium on Pollution, National Institute for Agricultural and Cattlery Research. Santiago, Chile: Instituto de Investigación Agropecuaria.

La Epoca. 1992. Article, August 9.

Lagos, Gustavo E. 1990. *Análisis de la situación del medio ambiente en realción con la minería Chilena*. Santiago, Chile: Centro para la Investigación y Planificación del Medio Ambiente (CIPMA).

Lagos, G. E., C. Noder, and J. Solari. 1991. *La situación Jurídica Institucional en el Area Minería y Medio Ambiente*. Santiago, Chile: Ministry of Mining.

Luna, R., and G. E. Lagos. 1990. *Evaluación del estado de las aguas del Río Aconcagua contaminadas por acción de origen minero*. Santiago, Chile: Centro para la Investigación y Planificación del Medio Ambiente (CIPMA).

————. 1992. "Metodología para el establecimiento de normas para efluentes metalúrgicos en la hoya hidrográfica de Copiapó." In *Proceedings of the Fourth Chilean Environmental Scientific Meeting*. Santiago, Chile: Centro para la Investigación y Planificación del Medio Ambiente (CIPMA).

Muñoz, L., and G. E. Lagos. 1990. *Análisis cuantitativo de la calidad de las aguas de la cuenca del Elqui*. Santiago, Chile: Centro para la Investigación y Planificación del Medio Ambiente (CIPMA).

Solari, J., and G. E. Lagos. 1991. "Strategy for the reduction of pollutant emissions from Chilean copper smelters." In *Pyrometallurgy of Copper*. New York: Pergamon Press.

Experimenting with Supranational Policies in Europe

DAVID HUMPHREYS

Policies for mining in Europe, including those relating to the environment, have historically been determined at the national level. However, among the countries of the European Community (EC), there has been a progressive tendency in recent years for more and more environmental policy to be determined at the supranational, Community level. This lecture looks at some of these policy developments and the potential impacts on the mining sector. It considers how the history of mining in Europe and the imperative of reconciling the needs and interests of twelve Member States appear to be pushing these policies in pragmatic and sometimes innovative directions.

First, however, it is necessary to give some perspective to the subject by considering the role played by mining in Europe and the manner in which EC law is framed and implemented. Although various references are made to Europe generally, the discussion is for the most part limited to Western Europe and, more particularly, to the twelve Member States of the European Community.[1]

---

[1]Belgium, Denmark, France, Germany, Greece, Ireland, Italy, Luxembourg, the Netherlands, Portugal, Spain, and the United Kingdom.

Europe's Mining Industry

Europe is not perhaps the first region one thinks of when the conversation turns to mining, yet Europe has a long tradition of mining as well as a significant mining industry today. This industry can be classified under three heads: the mining of metals, coal, and industrial minerals.

Metal mining has been carried out in Europe since the Bronze Age three to four thousand years ago. At the beginning of the last century, Europe (excluding, for present purposes, Russia) was mining half the world's copper and virtually all its lead and zinc. Even at the end of the century, it still accounted for roughly half the world's lead and three-quarters of its zinc though, by this time, only 17 percent of its copper (U.S. Bureau of Mines 1928, 1929a,b). Most of this production was focused on the United Kingdom (UK), Germany, and Spain. Although production tonnages were small by today's standards, little regard was paid to the environmental consequences of mining, and local impacts could be severe, as any visitor to Parys Mountain in Wales, to Rio Tinto in Spain, or the Harz Mountains in Germany can testify.

With the opening up of larger, richer resources elsewhere in the world, the importance of metal mining in Europe diminished and the region adapted to being an importer of ores and metals. Today, Europe accounts for less than 5 percent of the world's mine production of copper and tin, 10 percent of its lead, and 13 percent of its zinc. If one considers the EC alone, the figures are roughly half these. Against this, the EC accounts for about 30 percent of the world's consumption of nonferrous metals (World Bureau of Metal Statistics 1992). Although part of this metal deficit is made up from secondary (recycled) raw materials, it remains clear that Europe's domestic production potential stops well short of its metal consumption requirements.

One of the consequences of this is that, in contrast to most of the other regions covered in this volume, mining tends not to be as big a policy issue in Europe. Metal mining, as the highest profile type of mining and much the biggest generator of waste per tonne of product, simply does not have the same relative economic importance as in the other countries (see Table 1) and is thus less likely to be the subject of special measures. By the same token, where general policy measures do impact on the mining sector, there is rather less likelihood of the miner's voice being heard. The mining federations of Europe are not on the scale of the

Table 1. Comparison of Metal Mining Industries, 1991

Producer	Value of Output (US$ million)	Employment ('000)
European Community	744	14.2
United States	11,214	47.0
Australia	8,500	32.0

Sources: European Commission 1992; U.S. Bureau of Mines 1992; Australian Bureau of Statistics 1991.

Australian Mining Industry Council and the American Mining Congress. Europe is generally used to being an importer of metals, and strategic arguments for the defence of the domestic industry have little pull.

A qualification to this generalisation—particularly relevant to the policy context—is that the economic importance of metals mining tends to be greater in the southern states of Europe than in the northern ones. In 1990, Greece, Portugal, and Spain accounted for two-thirds of the total value-added contributed by this sector in the EC (European Commission 1992). As these are also among the countries with the lowest gross domestic product per head in the EC, it is easy to appreciate that the entry into the EC of these three states in the early 1980s added substantially to the policy interest in domestic mining.

The coal industry of Europe took off with the Industrial Revolution. Although output is now falling, Germany, Poland, Czechoslovakia, and the UK still rank among the world's ten largest producers (see Table 2).

Table 2. World Coal Producers, 1992

Producing Country	Amount produced (million tonnes)
China	1095
United States	907
Former Soviet Union	605
Germany	314
India	254
Australia	230
Poland	198
South Africa	175
Czechoslovakia	90
United Kingdom	87
Total World	4484

Source: International Energy Agency 1992.

The scale of this industry and its perceived strategic importance have long made coal the subject of special government provisions. In the UK, Poland, and Czechoslovakia, the coal industry has spent the last fifty years in state ownership—though in the UK, at least, this is soon to change. In what was East Germany, coal was similarly state-controlled, while in West Germany the industry was, and is, the subject of a special government-regulated subsidy—the so-called 'kohlepfennig' that power users are obliged to pay.

The importance of coal mining relative to other forms of mining means that for many in northern Europe, coal effectively *is* mining, and concern over the environmental impact of mining is focused largely, if not exclusively, on the impact of this sector. If one considers that, as far back as 1913, the UK and Germany were both producing almost 300 million tonnes of coal a year, these impacts—in the form of waste tips, runoff, subsidence, and unsightly abandoned workings—are clearly extensive. East Germany was mining and burning a similar quantity of high-sulphur brown coal, with even more devastating environmental consequences, just prior to reunification in 1990.

The third category of mining covers the industrial minerals used in such products as glass and ceramics, refractories, chemicals, paint, paper, and plastics, as well as the bulk minerals and materials used in construction. Although lacking the high profile of metals and coal, and often being the products of quarrying rather than mining, indus-trial minerals represent a substantial economic sector and one where Europe is largely self-sufficient. It is also a sector that faces many of the same environmental challenges as coal mining. The value of the output of this sector in the EC in 1991 is estimated at just over US$19 billion, an almost identical figure, as it happens, to that for the United States (European Commission 1992; U.S. Bureau of Mines 1992).

The Policy Context

In considering how EC policy is devised and implemented—particu-larly when comparing this process with that which applies in the other countries covered in this volume—it is critical to bear in mind that we are not dealing here with a single sovereign state but an amalgam of such states. The law-making process is consequently rather different and, needless to say, rather more complex.

At the centre of the EC law-making process stands the Commission of the European Communities. Although technically a bureaucracy—the EC's Civil Service—the Commission has certain powers not normally associated with bureaucracies. It has the right to initiate legislation, and it acts as a mediator between Member States where differences of opinion arise during the policy formation process. At the same time, its responsibilities for the execution of policies enacted by the Community are far more limited than would normally be expected of a bureaucracy.

If the Commission initiates and frames legislation, the Council of Ministers must eventually adopt it and is, in this sense, the Community's supreme legislative body. The Council is made up of representatives of the governments of the twelve Member States. Its composition varies according to subject matter. Thus, for foreign affairs, the Council comprises foreign minsters of the Member States and for environmental matters, environment ministers.

The manner of the Council's operation also varies according to subject matter. Until 1987 and the ratification of the Single European Act (SEA), all decisions had to be approved unanimously. However, facing the requirement to approve 282 pieces of legislation to bring about the Single Market in 1993, the Community partners decided to introduce the concept of majority, or 'qualified majority', voting over large swathes of policy, the principal exceptions being where there was an important sovereignty dimension, as for example with foreign policy or taxation. The SEA, by amending the Community's founding statute—the Treaty of Rome—made this possible. Under qualified majority voting, votes are weighted so as to give the larger states a greater influence on the outcome. Thus, the four largest states have ten votes; the smallest, Luxembourg, has two.

Laws enacted by the Community, of which there are two principal types, do not have quite the same meaning as laws enacted by the governments of sovereign states. The type of law that comes closest is the *regulation*. Regulations apply directly to all parts of the Community; they do not need to be transformed into domestic law. Like domestic law, however, they confer rights or impose duties and are required to be enforced by Member State governments. The second type of Community legislation is the *directive*. This is the principal vehicle for European legislation on the environment, as for most other policy areas. The directive is, as its name implies, a direction to Member States to enact, at a national level, such legislation as may be necessary to achieve a defined set of objectives within a specified period. Unlike the regulation, citizens only acquire the relevant rights and duties under a directive when it has passed into national law. (That said, it is possible for citizens to take their national governments to court for nonimplementation of directives.)

How a Member State incorporates a directive into national law is a matter for the Member State itself. The form of the directive derives from an acknowledgement that Member States are far better positioned than the Community institutions to assess the appropriate means of integrating the legislation with existing national law and adapting the legislation to special domestic circumstances. Not least among these special circumstances is the fact that EC Member States do not all share the same underlying systems of law. While the UK shares with the United States and the members of the Commonwealth a common law tradition rooted in the principles of pragmatism and precedent, most states in continental Europe have highly codified, rationalist systems following more directly the tradition of Roman law. This gives rise to differences in the formulation of law and, to the extent that UK law must be observed in the letter while Roman law requires only that the intention is there, in its enforcement. The advantages of having the Member States 'translate' European law into local terms are thus evident. It is, after all, the Member States that are required to implement it. The disadvantage, of course, is that it leads to variations in how states implement and enforce directives, a point addressed later.

The Environmental Dimension

There was no reference to the environment in the 1957 Treaty of Rome, and the Commission's early involvement in environmental legislation was justified on the grounds that differences in environmental standards in Member States obstructed the free flow of goods between them. Accordingly, most such legislation was pursued under that part of the Treaty of Rome (Article 100) that promotes the harmonisation of laws directly affecting the establishment of a common market (European Commission 1990).

The use of the legislative machinery in this way was in part a response to the reality that different parts of the Community wish to go at different speeds on environmental matters. The adoption by some Member States of stringent regulations on products and practices employed within their borders could have the effect of excluding the products of other countries. This remains a problem. A small, highly developed, densely populated country like the Netherlands, situated at the mouth of Europe's most heavily-exploited river (the Rhine), with a high water table and little available space for landfill, clearly faces a different set of environmental pressures to a country like Greece, which is sparsely populated and at a significantly earlier stage of economic development. Much Community legislation on the environment is in practice a transference of national legislation—notably that of Germany, the Netherlands, and Denmark—to the Community level in an attempt to keep intra-Community differences within tolerable limits. If not a level playing field, then at least the gradient needs to be controlled.

The Community's role in environmental matters was transformed by the SEA in 1987. This act formally recognised for the first time that environmental objectives need to take their place alongside the Community's other policies and, moreover, asserted that they should take as a base 'a high level of protection' (SEA, Article 100A.) To quote SEA's Article 130R, 'Action by the Community relating to the environment shall be based on the principles that preventative action should be taken, that environmental damage should as a priority be rectified at source, and that the polluter should pay. Environmental protection requirements shall be a component of the Community's other policies'. With the objectives came also the means to bring them about. By linking explicitly the environmental issue—more specifically, the issue of applying high, common standards across the Community—to the issue of trade liberalisation—specifically, the Single Market programme—considerable areas of environmental policy became the subject of qualified majority voting in the Council rather than, as previously, requiring unanimity.

The Treaty on European Union—the so-called Maastricht Treaty—will take this process one step further still. The concept of 'sustainable growth, respecting the environment' is incorporated into the basic principles of the Community, while Community competence is formally extended to the promotion of international-level environmental measures and policies 'based on the precautionary principle'. Substantial areas of environmental concern, not necessarily linked to the Community's economic objectives but to be pursued in their own right, become candidates for majority voting. Moreover, so as to ensure that the misgivings of the EC's poorer members do not unduly inhibit progress on environmental matters, the Treaty makes provisions for compensation from the Cohesion Fund when it is deemed compliance costs are 'disproportionate' for a Member State.

Another principle contained in the Maastricht Treaty that is of relevance to the future of environmental policymaking is that of *subsidiarity*. This is the principle—designed to limit any centralising tendencies within the Commission—that action should only be taken at a Community level in as far as the objective of the proposed action cannot be sufficiently achieved by the Member States themselves. Throughout the summer of 1992, against the background of attempts by member governments to get Maastricht ratified at a national level, the debate on subsidiarity, and what exactly it signified, raged. If inconclusive, it was nevertheless agreed at the Lisbon Summit not only that all future legislative proposals from Brussels would need to be justified against the principle of subsidiarity but—in a rider proposed by the UK—that certain *existing* EC legislation needed also to be reexamined in order to adapt it to the same principle. Significantly, the two examples of existing legislation advanced by the UK as illustrations of where the Commission had perhaps overstepped the mark, and unduly interfered in matters best dealt with at a national level, were environmental ones, specifically, directives on drinking water and on environmental assessments for land use decisions.

A counterpoint to developments in the institutional machinery of EC environmental policy has been an apparent gradual change in the philosophy underlying the policy itself. The early approach to environmental legislation tended to emphasise administrative instruments—licensing standards, emission limits, bans, and restrictions—an approach commonly termed 'command and control'. Legislation was introduced piecemeal, frequently in reaction to specific incidents or threats. The consequence of some of these legislative actions was, moreover, sometimes simply to transfer the problem from one medium to another, for example, from the land to the sea.

More recently, the approach has been evolving into something rather different. Greater emphasis is being accorded to an integrated approach to pollution control, one that seeks to eliminate environmental hazards at source and to promote a sustainable eco-balance in all media. There is a proposed directive on Integrated Pollution Prevention and Control, loosely based on 1990 UK legislation, which promotes the concept of 'best available technology' in the industrial field, and much talk in Brussels on the subject of 'eco-profiling', the environmental evaluation of products throughout their entire life cycles, from 'cradle to grave'. The approach is also more proactive than that taken in the past. Previously, EC legislation has, as mentioned, tended to follow that of the most environmentally progressive Member States. With the Council's approval in March 1993 of its Environmental Management and Audit Scheme regulation (6865/93), the Community, for the first time, began to lead Member States. At an international level, it sought at Rio de Janeiro in 1992, albeit unsuccessfully, to lead the world in the introduction of taxes on carbon dioxide emissions.

Another aspect of this change—most clearly evident in the EC's Fifth Action Programme adopted in May 1992—has been a shift in emphasis away from the legalistic, or administrative, model to a more pragmatic, cooperative one. There is a clearer acknowledgment of the value of economic instruments in achieving environmental objectives, of using market mechanisms rather than working against them. Principally through taxation, it is possible to change the structure of incentives to better reflect environmental costs and benefits. Eco-auditing can also be viewed as an attempt to make the market work for environmental objectives in as far as it is a means by which environmentally conscious companies can steal a public relations march on their rivals. There seems a readier acceptance that the pursuit of environmental objectives needs to be conducted on a cooperative basis between industry and governments within the context of a growing economy and not simply handed down from on high by governmental authorities irrespective of the costs involved. The environment is referred to as a 'shared responsibility', and far more emphasis than previously is given to the promotion of dialogue, voluntary agreements (an approach being pioneered by the Netherlands under its National Environmental Policy Plan), and other forms of self-regulation such as Responsible Care schemes. In short, in a programme the theme of which is, perhaps predictably, 'sustainable development', there seems to be a recognition that healthy, profitable businesses can play a more effective role in confronting environmental problems and promoting cleanup than ones bankrupted by the accumulation of insensitive environmental constraints.

Legislation Relevant to Mining

For mining, the most important direct effects of the EC's growing involvement in environmental matters are likely to arise from legislation on land management and on waste. The long history of mining in Europe, together with its relatively high population density, give these matters particular significance. In as far as solid wastes and tailings can give rise to problems of pollution from surface water runoff, in particular that which is acidic or contains heavy metals, then the EC's considerable battery of legislation on water also becomes relevant. Although likely to be less of a problem in Europe than in certain other parts of the world, concern over biodiversity could have some direct impacts on mining, particularly in the south of Europe where there is the greatest wealth of untouched wildlife and countryside.

In addition to the direct effects on mining, there could, of course, also be a host of indirect effects from EC environmental legislation. Legislation on emission levels at smelters and power installations, on secondary materials and their handling, on smelter wastes, on product risk and liability, on automobile standards and packaging, could all, in principle, have significant feedback effects on the demand for mined products. Indeed, these effects could, over the long run, prove rather more significant than the direct ones. As a massive consumer of energy, the metals processing industry will clearly be profoundly affected by any major changes in the value that society attributes to its energy resources. However, these matters are really the subject of a separate paper. The same goes for legislation on integrated pollution control and eco-auditing. Although both are important, they will have their initial impacts downstream of mining.

Land Use Planning

Land management is still a prerogative of EC Member States rather than the Community. However, there is one important way in which EC legislation impacts on the domestic decision-making process. This is the requirement, established under a 1985 Directive on Environmental Impact Assessment (85/337), that before authorising major developments with an impact on the natural environment, Member States' governments must ensure that a formal Environmental Impact Assessment (EIA) is carried out. Such assessments need to comprise an environmental baseline study and impact statement, and they are mandatory for certain specified projects, e.g., oil refineries and power stations. For the extractive industries, EIAs are carried out at the discretion of local or national authorities. Recent practice throughout the Community has been to subject new deep mines to EIAs and, also, open-cast workings of more than fifty hectares and smaller sites in sensitive locations (Johnson and Sides 1991).

To a degree, the directive simply confirmed what several Member States were doing, or proposed to do, anyway. However, the existence of the directive has given the Commission a locus in domestic land use planning decisions in as far as it can claim that proper procedures have not been followed prior to a decision on development being taken. A challenge by the Commission to the UK Government in the autumn of 1991, not over mining authorizations, as it happens, but over the adequacy of environmental impact assessments conducted ahead of several controversial transport schemes, led to some heated exchanges and helped set the stage for what was later to become the debate over subsidiarity. Under the Maastricht Treaty (Articles 130R and 130S), the Commission acquires new powers in respect of town and country planning matters in as far as these contribute to 'preserving, protecting and improving the quality of the environment' and to the 'prudent and rational utilization of natural resources', albeit that these powers are governed by unanimous voting.

Mine Wastes

The question of mine wastes has two distinct aspects: wastes generated as a product of existing and continuing operations and those that are the product of earlier, and frequently abandoned, activities.

Waste in the EC is governed by a framework Directive on Waste (75/442) introduced in 1975. This lays down the definition of waste, sets out general rules for waste disposal, and requires Member States to take measures to encourage waste prevention and re-use through recycling and reprocessing. Formally, mine wastes are excluded from the scope of the legislation where national legislation on mine wastes already exists (which, in most cases, it does). However, the somewhat contradictory reality is that some parts of the family of legislation on waste issues that fall under the framework do have direct relevance to mining. The framework directive has recently been overhauled with a view to raising standards and tightening conditions regulating operators in waste disposal and recycling, as well as aiming for a higher level of self-containment of waste within the EC as a whole and within Member States themselves. An amended Directive on Waste (91/156) was published in March 1991, with the requirement that its provisions be brought into law by Member States by April 1993.

Other relevant members of the family of EC waste legislation are directives on hazardous waste and a proposed directive on landfill. The former is essentially a classification of wastes according to the activity that generated them, their constituents, and their properties (e.g., toxicity). The original directive on the subject, the Directive on Toxic and Dangerous Waste (78/319), was introduced in 1978, though a new directive (91/689) was approved in 1991 and entered into force in Member States in December 1993. Six months prior to that, the Commission obtained agreement from Member States on a comprehensive list of hazardous wastes drawn up on the basis of the classifications contained in the directive. The list is consistent with that established in connection with legislation consequent to the Basel Convention on the transfrontier movement of wastes. Not unnaturally, the list contains a number of commonly used metals (e.g., cadmium and lead) and metal compounds (e.g., those of zinc and nickel).

The purpose of the proposed Directive on Landfill of Waste (COM(93)275) is to protect soil and water from pollution resulting from permanent land disposal of wastes. Landfills are to be divided into three categories: those for hazardous waste, those for inert waste, and those for a broad intermediate category of municipal, nonhazardous, and other compatible waste. While solid mine wastes, following the provisions of the framework directive, are excluded and do not need to be registered, leachates associated with these wastes are not excluded. Where effluent arising from mine wastes is deemed hazardous, the waste must be treated to reduce the extractability and thus make it suitable for disposal in a hazardous waste landfill. If the effluent cannot be satisfactorily treated, it must be disposed of in a 'monolandfill', that is, a landfill specific to the type of waste in question. What is inert or hazardous in this context will be determined by the Hazardous Waste Directive and by the specific provisions of the landfill directive itself. Thus, for example, to qualify as inert the total concentration of lead, cadmium, zinc, chromium, copper, nickel, and mercury in the leachate must be less than five milligrams per liter. To be designated hazardous, waste leachate concentrations will need to be 0.4–2.0 mg/l lead, 2–10 mg/l copper, 0.1–0.5 mg/l cadmium or chromium, and so on.

The standards proposed by the draft landfill directive are tougher than those applying in the United States and Canada, and it is believed that few operators will be able to meet them with existing process technologies (Riddler 1991). A further concern of the mining industry is that the directive appears to focus only on the composition of wastes, taking no account of their origin. While it may be the case that chemical treatment alters the solubility of mineral constituents, thereby potentially contributing to the degree of hazard, physical processing (e.g., crushing, grinding, gravity separation) does not. In this respect, runoff from the wastes generated by physical processing is little different from natural runoff. Indeed, to the extent that the processing may have removed certain constituents, including sometimes some of the more toxic ones, it could be considered less hazardous than natural runoff. To insist on the distinction is a variant on the claim that the Black Forest contravenes the EC directive on groundwater or that granitic rocks emit an unacceptably high level of radiation.

Liability and Historical Pollution

Regarding liability for wastes associated with past mining activity, the EC has, thus far, approached this matter with a certain circumspection. In the light of Europe's history, this is perhaps not surprising. As already noted, Europe's heyday as a metals miner was in the middle of the last century, and many of the mine sites concerned have long since been abandoned. The problems of identifying responsible parties, particularly those with the wherewithal to remedy the problem, would be not inconsiderable. The second point is that the extensive involvement of European governments in coal mining means that much of the responsibility for environmental cleanup lies with the government authorities themselves. In the UK, where the government is currently planning for the privatisation of the coal industry, there is no question of it seeking to pass on the environmental liability to prospective purchasers. Were the government to try, there would not be any purchasers. At a European level, government responsibility for the problem of the mining sector is reflected in the adoption in 1990 of the Rechar programme, under which financial support is available to certain regions to assist with the conversion of mining communities to other economic activities and to help restore former mine sites.

In 1989, the Commission produced a draft directive on Civil Liability for Damage Caused by Waste. Although heavily criticised—a symptom of the enormous difficulties of legislating in this field—the Commission persisted with the directive and in July 1991 published a revised draft. As in the United States, the Commission proposed a strict—as opposed to a fault-based—liability approach on the grounds that it would provide a stronger incentive for improved risk management in industry and that harmonisation of existing strict liability regimes already in operation in certain Member States was desirable on competition grounds. Still struggling to develop some sort of consensus on the issue, the Commission published, in May 1993, a further discussion document entitled 'Green Paper on Remedying Environmental Damage' (COM(93)47).

The amended draft directive, basing itself on the 'polluter-pays' principle, seeks to create a uniform system of liability to ensure that waste costs are internalised by industry and that victims of damage caused by waste receive fair compensation, as well as to establish a cleanup liability regime. It would hold producers jointly and severally liable for harm arising from their wastes. Where the producer cannot be identified, the person in 'actual control' (e.g., the landowner) would be deemed to be the producer. There is also a provision in the directive for mandatory pollution insurance.

Although this legislation would present many of the same kinds of problems for business as are already faced in the United States, the proposed legislation does not, on the face of it, imply retroactivity—i.e., liability for past as well as current wastes—as does U.S. Superfund legislation. Certainly, this is the hope and expectation of the mining industry, and it would seem to represent a stance consistent with the principles expressed in the Fifth Action Programme, where the emphasis is on the development of workable solutions to environmental problems rather than on the attribution of blame. The U.S. experience has been a sobering one, and it must be doubted whether EC Member States have the political will to go down this road, particularly since the countries that might otherwise be most disposed to this approach are also the ones where the historical mining problems are most acute.

That said, there are few grounds for complacency in the industry. For a start, the draft directive is not wholly unambiguous on the matter of retroactivity. Liability, it states, is to apply only to harms arising from an 'incident' occurring after the directive's implementation. However, there is a problem over the definition of an 'incident'. In as far as an incident occurring after implementation could be attributed to waste generated before that time—long-term leaching is a case in point—it is by no means clear, as currently drafted, that the directive does wholly preclude retroactivity (Smith and Hunter 1991).

There is also the question of the alternative arrangements that the EC may come up with to deal with historical pollution. The Commission's green paper advances the idea of 'joint compensation systems' to apply in circumstances where it is difficult to apportion liability for damages caused by past activities, releases from multiple sources, or by an unknown polluter. In the words of the green paper, this 'would spread the costs of restorative actions among specific economic sectors or geographic regions. The aim would be to pitch responsibility to pay for the costs of environmental damage close to the source, by aiming at the lowest possible level of activity linked to the damage' (Environmental Data Services 1992a). In short, while individual mine operators might escape liability for the sins of their forefathers, the industry may not.

Enforcement

One final area that deserves mention, not least because it highlights another of the differences in approach to environmental matters evident in the United States and Europe, is that of enforcement.

The European Commission has demonstrated itself highly effective at initiating legislation, building a consensus around it, and gaining approval for it in the Council. However, it has been far less effective at seeing to its implementation. Certainly it has nothing available to it approaching the legal powers of enforcement that the Environmental Protection Agency (EPA), Occupational Safety and Health Administration, or Mine Safety and Health Administration possess in the United States. Although this arises in part from the different underlying approach of the EC to environmental affairs, it also, of course, reflects real differences in the political structures concerned. As constituted, implementation of EC law is primarily a responsibility of Member States themselves and the zealousness with which states approach this responsibility varies. Although the Commission has certain powers to call recalcitrant members to order, its monitoring system is weak and heavily dependent on public complaints procedures.

In 1991, more complaints (353) were lodged in connection with alleged breaches of environment law than with any other area of EC law. The environment was also in first place in the league table of enforcement actions initiated by the Commission. During the year, 136 letters of formal notice—the first stage in the official infringement procedure—were sent out by the Commission. In addition, there were fifty-four 'reasoned opinions'—the penultimate stage in the infraction process—delivered, and eight cases referred to the European Court

(Environmental Data Services 1992b). The same report from which these figures were taken also showed the records of Member States, up to the end of 1991, on implementation of some ninety or so directives in the environmental field that were due to have been implemented by that date, as well as references to the European Court over 1987–1991 for alleged breaches of EC environmental and consumer protection laws. (These are reproduced in Table 3.)

Table 3. Environmental Directives Implemented and Alleged Breaches of Environmental Directives Referred to European Court: Record of Member States of the European Community

	Environmental Directives Implemented (%)	Alleged Breaches Referred to Court
Belgium	81	9
Denmark	98	–
Germany	92	7
Greece	76	5
Spain	93	3
France	89	10
Ireland	84	4
Italy	59	16
Luxembourg	86	2
Netherlands	95	3
Portugal	94	–
United Kingdom	85	2

Source: Environmental Data Services 1992b.

Tabulations of this sort are all part of the peer group pressure used to force EC states into compliance. It is a system that has its attractions but also its limitations, not least because enforcement procedures generally have to be triggered by formal complaints and the citizens of different countries vary considerably in their expectations of government and in their disposition to complain. Frustration over what is seen as foot-dragging by certain Member States, and particularly where this foot-dragging is perceived to give competitive advantages, has led other members to press for improved implementation measures.

It is partly to address this concern that the Community has decided to set up the European Environment Agency (EEA). The remit of the EEA is 'to record, collate, and assess data on the state of the environment, to draw up expert reports on the quality, sensitivity, and pressures on the environment within the territories of the Community' and 'to provide uniform assessment criteria for environmental data to be applied in all member states' (Berbotto 1991). The intention is that the role of the agency will be essentially advisory and information gathering. Unlike the EPA in the United States, EEA will have no executive authority. Although the EEA should arm the Commission better in operating its infraction procedures and give greater authority to its judgments, the EEA essentially represents an affirmation of the principle that simply by bringing certain information to the public's attention, particularly that on comparative performance, governments can be shamed and bullied into taking remedial action. In a sense, the system seeks to work by mobilising public opinion over the heads of Member States' governments. The same principle underlies a directive on public access to environmental information, approved in 1991, which will oblige member governments to divulge information that they hold on environmental performance on request, except that where to do so would be contrary to the national interest.

This said, there are many who believe the establishment of the EEA is simply the thin end of the wedge and that the EC is headed inexorably down the same road as the United States, a view which the Commission has done little to discourage. The possibility of a stronger role for the EEA in the future seems to be contained in the founding regulation that says that the agency can 'associate in monitoring the implementation of Community environmental legislation'. Recently, the idea of establishing a European Environmental Audit Inspectorate has been garnering support. It is envisaged that such an inspectorate, working in tandem with the EEA, would coordinate and audit national inspectorates, thereby ensuring that EC environmental directives and regulations are consistently and effectively implemented by Member States. The Maastricht Treaty provides further formal powers of enforcement for EC legislation by giving the European Court of Justice power to fine Member States for noncompliance with its rulings.

Concluding Remarks

European legislation for the environment has thus far had only a limited impact on mining. This reflects essentially two things. First, the relatively small scale of Europe's metal mining sector and the relatively localised effects of mining's environmental impacts have meant that this has not been a priority target for the Commission's environmental enthusiasms. Second, it reflects the reality that the EC's formal powers on environmental matters are of relatively recent origin and that the major impacts of environmental legislation for the mining industry still arise principally at the national level.

However things are changing. This lecture has pointed out the explosion of environmental legislation emanating from Brussels since the Single European Act of 1987 and the shift in the balance of environmental jurisdiction from the national to the supranational level. The Maastricht Treaty, finally ratified in October 1993, will extend the powers of the Community institutions in the environmental area still further. The precise impacts are uncertain but are likely to be most prominent in preproduction environmental assessments and in the treatment of mining wastes. And, as everywhere, tightening standards will be associated with tougher operating conditions and increased costs, particularly front-end costs. There is the additional problem, touched on in the discussion on wastes, that the mining industry's diminishing size in Europe, and a consequent lack of knowledge about mining within government circles, carry increasing risks that the industry will be subject to legislation that has been devised with other industries in mind and takes insufficient account of mining's particular characteristics. There is, in a word, a problem of mining's critical mass.

It needs to be stressed, however, that increasingly stringent environmental conditions are only one of the several difficulties confronting European mining. Other important factors include the impact of economic recession on demand, resource exhaustion, a declining commitment by government to support coal mining, the knock-on effects of this on mining education, and scepticism among the European public about the need for mining—or at least the need for it in their overpopulated backyard. These impacts are cumulative, and tightening environmental legislation enters the equation as simply one more problem to be confronted and one more cost to be borne. But then again, it is, more obviously than the other factors, a discretionary matter for government and therefore a more obvious target for complaint.

Working against the centralising tendencies in environmental legislation is the countervailing force of subsidiarity. The summer of 1992 saw a backlash, notably in the UK and Denmark (which rejected Maastricht in its first referendum), against what was perceived as the undue interference of Brussels in national matters. The principal environmental consequences of mining being essentially local ones, the sector does not immediately suggest itself as a natural candidate for unified legislation. Certainly the case is a different one from that of airborne emissions, which have clear cross-border implications. The principle of subsidiarity, in this context at least, also finds support in the countries in the south of Europe, where mining plays a more prominent role economically and which tend to have a different set of development priorities to the richer countries to the north. Tacit evidence of the Commission's acceptance of this reality was provided in September 1992, when it approved a derogation permitting the Spanish government to give financial support to its mining industry at the same time that the Commission was enjoining the German government to reduce its support for coal mining.

Alongside the developing institutional arrangements governing environmental policy, the underlying philosophy of the EC's approach to environmental questions appears also to be changing. Although early EC forays into environmental matters tended to be of the 'command-and-control' variety, more recently—and most explicitly in the Fifth Action Programme—there has been a shift away from the more formal, legalistic approach toward one that is more cooperative and pragmatic in character and that envisages the use of a wider range of policy instruments. This approach has been forced on the Community partly by the political reality of having to reconcile the interests and aspirations of twelve separate states, but also perhaps partly by an increasing awareness of the limitations of the legalistic approach as demonstrated by the U.S. experience.

The jury is still out as regards both the durability of this change and its likely implications for individual industries. However, the language of environment as a 'shared responsibility' is encouraging, as is the explicit acknowledgement that the objectives of environment and development are not inherently opposed. Quite how 'voluntary' voluntary measures for improving environmental behaviour will prove to be will have to be seen. On the matter of cleanup, though the industry is by no means out of the woods, there are signs of an acceptance that the real challenge lies in finding practical ways to tackle environmental problems, not in finding guilty parties. The assumption by governments of part of the financial liability for the legacy of the past may permit these solutions to be arrived at the faster. After all, society as a whole enjoyed the benefits of cheap raw materials in the past, and it is not unreasonable that society as a whole should contribute towards the restitution of these problems.

Subject to the success or otherwise of Europe's emerging policy thrust, there may be lessons for others' attempts at supranational policymaking, and perhaps national policymaking also. The development of policies through a slow and low-profile process of technical and political negotiation acknowledges that the environment is not generally a matter for absolutes but for trade-offs between numerous social and economic choices and should provide for a more thorough integration of the relevant variables than is possible under a more legalistic approach. Moreover, the process of negotiation itself is one of consensus-building, leading participants gradually toward solutions that can be seen to be achievable, while at the same time generating obligations for the enforcement of these solutions.

No less important, unlike strict regulatory approaches, which by their nature need to be applied equally everywhere, the more political approach to environmental matters permits recognition that not all countries share the same development aspirations and allows policies to be moderated accordingly. Where no cross-border implications exist, this is readily achieved by leaving the matter largely in the hands of national authorities, though perhaps within the framework of broadly shared policy objectives. Where there are cross-border implications, as will generally be the case with supranational policymaking, then exceptions and compensation arrangements may have to be accepted as a necessary price for obtaining an improvement overall. The most recent proposals on energy and carbon tax from the Commission make provision for a partial exemption for the southern states of Europe in recognition of their specific development requirements, and generally it is coming to be accepted that global emission limitation schemes that do not contain provisions of this nature have no chance whatever of being acceptable politically. Every bit as important as what is theoretically desirable in environmental affairs is what is practically achievable.

The future impact of EC legislation on mining can thus be seen as part of a much broader picture that is itself only gradually coming into focus. The debate on subsidiarity is helping to define the appropriate scope of Community action in this and in other spheres. The evident difficulties with too legalistic an approach are leading to the elaboration of broader and more flexible policy instruments. The failure of the Commission to push through its ill-prepared plans for an energy and carbon tax ahead of the Rio Summit in 1992 should have helped persuade it conclusively that the environment is not an appropriate subject for heroic gestures, but is rather something that needs to be the subject of detailed and laborious analytical work relating to risk as well as hazard and to the objective assessment of costs and benefits of actions proposed. There is some evidence in the Commission of a new breed of realism and of practitioners who talk in these terms rather than in the missionary language of environmental absolutes. Environmental standards for the mining industry in Europe, as for most other industries, will undoubtedly get tougher with time, but it will help if in the elaboration of those standards it is possible to employ a common language and a common framework of values.

Acknowledgements

The assistance of Jim Stevenson, RTZ Group Environmental Scientist, and Dr John Bramley, Managing Director, Laporte Minerals, is gratefully acknowledged.

References

Australian Bureau of Statistics. 1991. *Yearbook Australia 1991*. Canberra: Commonwealth of Australia.

———. 1992. *Mineral Production Australia 1990–1991*. Canberra: Commonwealth of Australia.

Berbotto, Paolo. 1991. EC Waits for Its Environment Agency to Reveal Its Strengths. ECN *Environment Review Supplement* (July): 18–20.

Environmental Data Services. 1992a. ENDS *Report*. no. 204 (January).

———. 1992b. ENDS *Report*. no. 213 (October).

European Commission. 1990. *Environmental Policy in the European Community*. 4th ed. Luxembourg: Office for Official Publications of the European Communities.

———. 1992. *Panorama of EC Industry*. Luxembourg: Office for Official Publications of the European Communities.

International Energy Agency. 1992. *Coal Information 1992*. Paris: Organization for Economic Co-operation and Development.

Johnson, M. S., and A. Sides. 1991. Environmental Assessment of New Mining Projects. In *Minerals Industry International*, 13–17. London: Institution of Mining and Metallurgy.

Riddler, Gordon P. 1991. Exploration Strategy and the Natural Environment: From Grassroots to Jumbos. In *Minerals Industry International*, 8–12. London: Institution of Mining and Metallurgy.

Smith Jr., Turner T., and Roszell D. Hunter. 1991. The Revised European Community Civil Liability for Damage from Waste Proposal. *Environmental Law Institute's Reporter* (December).

U.S. Bureau of Mines. 1928. *Summarized Data of Copper Production*. Washington, D.C.: U.S. Government Printing Office.

———. 1929a. *Summarized Data of Lead Production*. Washington, D.C.: U.S. Government Printing Office.

———. 1929b. *Summarized Data of Zinc Production*. Washington, D.C.: U.S. Government Printing Office.

———. 1992. *Mineral Commodity Summaries 1992*. Washington, D.C.: U.S. Government Printing Office.

World Bureau of Metal Statistics. 1992. *World Metal Statistics*. Ware, United Kingdom: World Bureau of Metal Statistics.

The Limitations of Environmental Regulation in Mining

ALYSON WARHURST

Public policy to promote technical change and foster economic efficiency, in addition to environmental regulation, is most likely to contribute towards sustained and competitive improvement in the long-term environmental management of nonrenewable natural resources. Three main issues are analysed in this lecture, drawing out implications for the different contexts of both industrialised and developing countries.

The first issue is the relationship between production efficiency and environmental performance. There is growing evidence that technical change, stimulated by the 'Environmental Imperative', is reducing both production and environmental costs to the advantage of those dynamic firms that have the competence and resources to innovate. Such firms include mining enterprises in developing countries as well as transnational firms, but the evidence is strongest for large new investment projects and greenfield sites. In older, ongoing operations, environmental performance correlates closely with production efficiency. Environmental degradation is greatest in operations working with obsolete technology, limited capital, and poor human resource

The author would like to acknowledge gratefully the kind assistance of Sarah Hannant, Gill Partridge, and Hilary Webb in the preparation of this paper. Support for the research reported in this paper was provided by the John D. and Catherine T. MacArthur Foundation. Parts of this paper build on a more detailed study (see Warhurst 1994).

133

management. Policy to promote the development of the technological and managerial capabilities to effect technical change in those organisations would lead to improved efficiencies in the use of energy and chemical reagents and in waste disposal; to higher metal recovery levels; and to better workplace health and safety. This in turn would result in improved overall environmental management.

The second issue is the economic and environmental limitations of regulation. Currently, the environmental performance of a mining enterprise is more closely related to its capacity to innovate than to the regulatory regime within which it operates. Although international standards and stricter environmental regulation may not pose problems for the economics of new mineral projects, there could be significant costs involved for older, and particularly inefficient, ongoing operations. Controlling pollution problems in many of these cases requires costly add-on solutions: water treatment plants, strengthening and rebuilding tailings dams, scrubbers and dust precipitators, etc. Furthermore, in the absence of technological and managerial capabilities, there is no guarantee that such items of pollution control—environmental hardware—will be incorporated or operated effectively in the production process. In some instances, such requirements are leading to shutdowns, delays, and cancellations, as well as reduced competitiveness. When mines and facilities are shut down, the cleanup costs frequently get transferred to the public sector, which—particularly in developing countries—has neither the resources nor technical capacity to deal with the problem effectively.[1]

The third main issue is the case for an environmental management policy. The implication of this analysis is that to ensure competitive and sustainable environmental management practices in metals production, governments need to embrace public policy that goes beyond traditional, incremental, and punitive environmental regulation. The latter, in the old 'environmental protectionist' mode, tends to treat the symptoms of environmental mismanagement—pollution—and not the

---

[1]The Superfund legislation of the United States enables blame for environmental damage to be apportioned to a selected one of many past mine owners and for that company to be charged an estimated cost for work that government contracts to clean up and rehabilitate the damaged site. In most countries, perhaps with the exception of the United States, the lack of retroactive regulation means the pollutee-suffers-and-pays principle is alive and well. This is not in itself, however, an argument either against regulation per se or for the global diffusion of Superfund legislation.

causes—lack of capital, skills, and technology and the absence of the capability to innovate. The challenge will be for governments to ensure that firms operating within their national boundaries remain sufficiently dynamic to be able to afford to clean up when operations cease and to innovate to improve economic efficiency and environmental management in the meantime.

Governments need policy tools that enable them to predict the warning signs of declining competitiveness and impending mine shutdown to ensure sufficient resources are available for the environmental management of mine 'decommissioning'. Policy mechanisms need to be developed that promote technical change and build up the technological and management capabilities to innovate and manage the acquisition and absorption of clean technology. The privatisation of the state sector and the liberalisation of investment regimes in many developing countries (such as Angola, Mozambique, Namibia, Botswana, Bolivia, Peru, and Chile), with their emerging emphasis on joint ventures and interfirm collaborative arrangements, provide new opportunities for the diffusion of both competitive and environmentally sound best-practice in metals production.

Public policy to promote technical change and, complementing that, to improve economic efficiency respects the interplay between the environmental and economic factors that constitute a sustainable development approach to the long-term environmental management of our nonrenewable natural resources. Environmental regulation at best provides only one element of a public policy for environmental management.

The Nature of the Environmental Problem

Mining and mineral processing produce waste products and ecological disruption, which may generate potential environmental hazards at each stage of the metal production process. Figure 1 depicts some of these waste products and their associated hazards, which have local, regional, and, to a lesser extent, global manifestations. Mining-related activities affect all three environmental media—land, water, and air. Exploration, mine development, and the dumping of barren overburden or waste can degrade the habitats of local flora and fauna and prohibit alternative land-uses—forestry, agriculture, or leisure. Land effects include land-use conflicts, particularly over sources of energy (such as forests) for mineral processing, and ecosystem disruption.

135

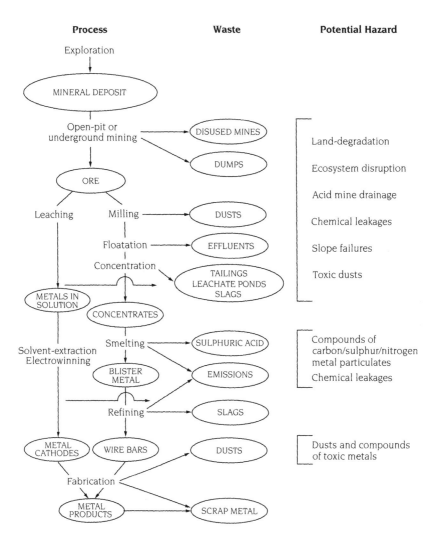

Figure 1. The mining process and the environment: processes, waste, and potential hazards are interrelated.

Source: Warhurst 1993

Potentially toxic chemicals—xanthates—are used in primary processing, as are inorganic reagents such as zinc sulphate, copper sulphate, sodium cyanide, and sodium dichromate. The tailings—left by river banks and often in poorly secured ponds—contain residuals of both toxic metals and chemical flotation agents.

Water quality may be affected by naturally occurring acid mine drainage from mines and waste piles. Certain ores contain considerable quantities of sulphides, which, through oxidation and naturally occurring biological processes, are transformed into sulphates and sulphuric acid, which in turn leach out further metal values from adjacent waste or ore. Water draining from surface mines, wastewaters from processing, and rainwater runoff all require careful monitoring and control, which are rarely undertaken systematically in developing countries, with disastrous results. By their very nature, mine water pollution problems, once they have begun, can only be treated retroactively—if at all—at excessive cost. The oxidation process that creates acid mine drainage is autogenous, and there is no end to the volume of sulphide waste and rock to 'feed' it.

Toxic leaks and spillages from tailings dams or reagent ponds may also affect water quality. This applies to the inadequate management of toxic wastes from the hydrometallurgical treatment of gold ores with cyanide and to semiliquid wastes from the concentration and smelting of nonferrous metals, including copper, tin, lead, zinc, silver, and nickel. The red mud slag produced by bauxite mining can also be categorised as potentially hazardous in this respect. Significantly, many of these water quality effects can cause environmental hazards beyond the life of the mine.

Air quality may also be affected by emissions from roasters and smelters, and to a lesser extent from refineries, of compounds of carbon, sulphur, and nitrogen and of toxic metal particulates. There are also indirect emission effects from the use of fossil fuels as energy and potentially hazardous dusts and gases released in the workplace.

It is a complex task to estimate the costs of natural resource degradation associated with mineral exploitation and to design public policy to share those costs among the polluter, state, and community. Increasingly, it is becoming apparent that these costs are high. In the past, the costs have largely been measured in terms of the expense involved in, for example, remedial treatment of degraded water quality, investment in environmental control technologies, or compensation for damage by toxic dust to local farmland. More recently, particularly in North America, environmental costs have been estimated in terms of extensive rehabilitation programmes, which transform the previous mine and plant site for alternative resource uses, such as revegetation or leisure facilities. Indeed, in the context of developing countries, one could argue that the mining industry was traditionally structured to externalise such environmental costs so that the maximisation of profit was achieved not so much through efficiency and innovation, but through the appropriation of undervalued resources and the shifting of the environmental costs of doing so on to others.[2]

Although some environmental degradation is the inevitable result of mining, examples can be given where pollution either has negative economic impacts or presents economic opportunities—for firms as well as governments. For example, by-products that are toxic but could warrant economic reprocessing are frequently dumped. This is especially the case in developing countries, where inaccurate sampling or inefficient technologies result in such loss. Similarly the mining of high-grade ore and the dumping of low-grade ore, which is a short-term expediency to boost foreign exchange earnings in times of crisis, result in greater environmental degradation (higher risk of acid mine drainage from dumps) and the loss of longer term revenue. Costly water treatment projects are often instigated as part of the mine closure programme rather than implementing acid mine drainage prevention from the outset of the mining project, involving much cheaper pollution control and often resulting in the recovery of metal values. Finally, some firms have been obliged to pay the health care costs of communities as a result of their drinking degraded water, costs which in many cases outweigh the cost of technical change to treat the chemical effluents in the first place.

There is however still considerable work to be undertaken to quantify the nature and extent of environmental degradation caused by metals production. Currently there exist only isolated case studies and little systematic analysis of the problem. It is difficult to generalise since local geology, geography, and climate affect mineral and ore chemistry,

---

[2]When it comes to evaluating these costs, it should be remembered that those most affected by environmental pollution from mining in developing countries are generally those least able to understand and respond to it—remote miners' families or isolated rural communities—or they respond in a short-term, unsustainable way. That might mean, for example, in the case of peasants whose farmlands were ruined through pollution from the Karachipampa tin volatilisation plant in Bolivia, that small compensation payments were offered to cover only the loss of one particular harvest rather than the potential loss of their livelihoods. In contrast, in the United States, propelled by Superfund laws, the assessment of natural resource damage liability is a fast-growing area of consultancy (Kopp and Smith 1989).

soil vulnerability, and drainage patterns, and hence the extent of environmental hazard created. Furthermore, a major factor affecting whether a hazard results in environmental degradation is the social and economic organisation of the production unit, including its size, history, and ownership structure, as well as its propensity to innovate.

Limitations of Environmental Regulation and the Challenge for Public Policy

Regulatory Effectiveness in Industrialised Countries

Regulatory frameworks for safeguarding the quality and availability of land, water, and air that might be degraded as a result of mining and mineral processing activities are growing in number and complexity. This has particularly been the case in the major mineral-producing countries of North America and Australia, as well as Japan and Europe. The norm in environmental regulation is that governments set maximum permissible discharge levels or minimum levels of acceptable environmental quality. Such command-and-control mechanisms include: Best Available Technology standards, clean water and air acts, Superfunds for cleanup and liability determination, and a range of site-specific permitting procedures, which tend to be the responsibility of local government within nationally approved regulatory regimes. Command-and-control mechanisms tend to rely on administrative agencies and judicial systems for enforcement. Three issues are relevant regarding the appropriateness of command-and-control environmental regulations for reducing environmental degradation and improving environmental management practices in metals production.

First, there is a trend away from a 'pollutee-suffers' to a 'polluter-pays' principle. However, it remains the case that the polluter pays only if discovered and prosecuted, which requires technical skills and a sophisticated judicial system, and that this occurs only after the pollution problem has become apparent and caused potentially irreversible damage. This highlights the tendency of such environmental regulations to deal with the symptoms of environmental mismanagement (pollution) rather than its causes (economic constraints, technical constraints, lack of access to technology or information about better environmental management practices). This can be serious in some instances, because once certain types of pollution, such as acid mine drainage, have been identified, it is extremely costly and sometimes technically impossible to trace the cause, rectify the problem, and prevent its recurrence. Certain environmental controls may only work if incorporated into a project from the outset (e.g., buffer zones to protect against leaks under multitonnage leach pads and tailings ponds).

Second, Best Available Technology (BAT) standards may be appropriate at plant start-up, but their specified effluent and emission levels are not necessarily achievable throughout the life of the plant, because technical problems may arise and there may be variations in the quality of concentrate or smelter feed, etc., if supply sources are changed. Moreover, there are serious implications for monitoring. It would also be erroneous for a regulatory authority to assume standards are being met if a preselected item of technology has been installed. Ongoing management and the environmental practices at the plant are also likely to be important determinants of best environmental practices.

Third, related to points one and two above, BAT standards and environmental regulations of the command-and-control type tend to presume a static technology—a best technology at any one time. This tends to promote incremental, add-on controls to respond to evolving regulation rather than to stimulate innovation. This acts as a disincentive to innovate by equipment suppliers, mining firms, and metal producers. Their innovation, which has required substantial R&D resources, may be superseded by some regulatory authority's decision about what constitutes BAT for their particular activity. BAT gives the impression of technology being imposed from outside the firm, not generated from within. The search for profit and cost-savings tends to be a more obvious instigating factor of technical change, and it might be argued that market-based mechanisms, a technology policy that is complemented by a regulatory framework, and a good corporate environmental management strategy can better contribute to achieving that aim.

There has been growing interest in the use of market-based mechanisms, whereby the polluter is charged for destructive use by estimating the damage caused. An important justification for the use of market-based incentives is that they allow firms greater freedom to choose how best to attain a given environmental standard (OECD 1991). By remedying market failures or creating new markets (rather than by substituting government regulations for imperfectly functioning markets), it has been argued, market-based incentives may permit more economically efficient solutions to environmental problems. Two categories of incentives exist (O'Connor 1991; Warhurst and MacDonnell 1992). One set, based on prices, includes a variety of pollution taxes, emission charges, product charges, and deposit-refund systems. Another set is quantity-based and includes tradeable pollution rights or marketable pollution permits. The most common of these measures relates to posting bonds up front for the rehabilitation of mines following closure. This is standard practice now in Canada and Malaysia. There are also discussions taking place about a mercury tax in Brazil and a cyanide tax in the United States. Currently, no government has designed a systematic set of incentives for industry to innovate and develop new environmental technology.

There are two further areas where policy approaches can contribute to improved environmental management practices. The first approach increases the use of conditionality in the provision of private, bilateral, and multilateral credit. This approach frequently requires both prior environmental impact assessment and the use of best-practice environmental control technologies in new mineral projects. A growing number of donor agencies in Germany, Canada, Finland, and Japan, for example, are also concerned with training in environmental management. The second approach encompasses the attempts by some governments, particularly Canada, to stimulate R&D activities (jointly and within industry and academic institutions) to determine toxicity from mining pollution and to promote cleanup solutions. For example, Canada has extensive government-funded R&D programmes to promote the abatement of acid mine drainage and of sulphur dioxide emissions. There is considerable scope for expanding these approaches, as will be argued below.

Regulatory Effectiveness in Developing Countries

Environmental regulations designed specifically for mining and mineral processing have until recently been uncommon in developing countries, although most countries now have in place basic standards for water quality and, less commonly, air quality. A few developing countries have recently adopted extensive regulatory frameworks—sometimes replicas of U.S. models. This, for example, has been the case in Chile and, to a lesser extent, in Brazil. This growing concern about environmental degradation is occurring during a period of rapid liberalisation in developing countries (Brown and Daniel 1991), which finds expression in new policies to promote foreign investment, privatisation schemes, and the availability of loan capital. These conditions also influence the regulatory regime of developing countries. Should the developing country pose less onerous environmental burdens on the potential investor to improve the terms of the investment by implying lower compliance costs or a greater assumption by the state of the environmental costs associated with mineral development projects? Should agreements be signed that release new investors from any liability for environmental damage caused by previous mine owners under less restrictive regulatory regimes? Or will a clear and strict regulatory regime be more likely to facilitate credit flows from increasingly more environmentally-conscious lending agencies? Developing countries desperate for investment in their stricken mineral sectors will need to determine what the market can stand and how such terms can be structured to reduce to the minimum the risk premium the investor will seek for a given tax or regulatory burden.

It is worth noting that a survey by Johnson (1990) and work by Eggert (1992) imply that environmental policy has not been a major factor in determining the investment strategies of international mining firms. However, more recently, the industry press has been citing environmental regulations in Canada and Australia as a major factor causing the cancellation and delay of potentially large investment projects (*Mining Journal* 1992) and contributing to the shutdown of several mines. For example, in 1989, the Bharat Aluminium Company announced the closure of its bauxite mining project in the Gandhamardham Hills, Orissa State, in India, because of strong environmental opposition by the local population (U.S. Bureau of Mines 1989, 6). Other projects, such as the Phelps Dodge Copper Basin project in Yavapai County (United States), have been withdrawn due to delays and excessive costs involved in project approval, while in 1991 the Kennecott Flambeau Mining Company finally received planning permits after twenty years of negotiation for the Grant Copper Mine in Wisconsin, which will operate for only six years.

However, environmental regulation alone is unlikely to solve environmental problems in developing countries because of endemic production inefficiencies. In particular, the approach of state-owned enterprises toward the environment reflects inefficient operating regimes, excess capacity, breakdowns and shutdowns, and poor management procedures, which contribute to worsen pollution. Such inefficiencies make it very unlikely that environmental controls will be incorporated effectively. Moreover, obsolete technology is widely used without the necessary modern environmental controls and safeguards. For example, new concentrators and roasting plants tend to be totally computerised. Automatic ore-assaying techniques give an extremely accurate picture of the chemical composition of the ore feed, with implications for the fine-tuning of pressure, heat, cooling, and specific environmental control systems. This in turn enables the accurate prediction and monitoring of emissions. However, where these controls are missing and, in particular, where ore feeds are of variable composition (in terms of the sulphur, lead, and arsenic content), emissions also vary with regard to their pollutant content. The inefficient use of energy and poor energy conservation practices also result indirectly in increased environmental pollution through the excessive burning of fossil fuels. This is particularly the case in poorly lagged roasters and inefficiently operated flotation units and smelters, which are very intensive in energy use. It might be further argued that command-and-control regulatory instruments are unlikely to result in a reduction of pollution since they cannot affect the capacity to implement technical change of a debt-ridden, obsolete, and stricken mining enterprise in the developing country context. Such a firm might find it preferable to risk not being detected or convicted, to pay a fine, or to mask its emission levels, rather than face bankruptcy through investing in radical technological change.

In addition to the problems of inefficient production, there are a range of further reasons why environmental regulations—particularly those of the command-and-control, incremental, and cumulative nature (Panayotou, Leepowpanth, and Intarapravich 1990)—do not improve environmental management, particularly in developing countries. These reasons are discussed below.

Environmental regulations tend to be of the blanket-type, which specify maximum levels of emitted substances, minimum levels of environmental quality, and best available technology standards. They do not tend to reflect the propensity of a particular operation to pollute, which in part depends on local site-specific conditions (geology, geography, and climate) as well as economic, infrastructure, and technology-related constraints. In a desert, tailings dams need not be as highly specified as in rainy climates; dust regulations may need to vary depending on topography, precipitation, and prevalent winds; the substrata of leach ponds might need to be of different composition, strength, and depth, depending on local geology or the existence of an impermeable level of clay. Since developing country regulations are often copied directly from the statute books of industrialised countries, they may not be appropriate for the site-specific characteristics of mines in either tropical regions or deserts. They may result in unnecessary and costly adaptations on the one hand, or the lack of necessary controls on the other.

Command-and-control environmental regulations require intensive monitoring to ensure that they are enforced. However, the small and medium mine sector accounts for at least 25 percent of mineral production in many countries. Although these mines are individually relatively small polluters, collectively they account for a disproportionately large share. These mines are often located in high mountains, in remote tropical rain forests, or in politically dangerous regions, and they are almost impossible to monitor systematically. Indeed, as regulation becomes more sophisticated, such monitoring requires skills and human resources far beyond the technological and managerial capabilities of many developing countries and frequently beyond their budgets. Understanding the diverse range of toxicity and engineering issues behind regulatory aims also poses challenges even in the industrialised countries, and the most knowledgeable regulators are often head-hunted by the mining firms.

Moreover, the enforcement of command-and-control regulations depends on a system that admonishes with imprisonment and fines. This in turn requires a legal structure and judicial system far beyond the capacity of most developing countries. Compliance is also limited, since fines are generally a fraction of the costs involved in remedial treatment and abatement technology. They are only payable if the polluter is detected and convicted. Inflation and local currency devaluation, which are endemic in developing countries, also eat into the value of such fines. The costs of environmental regulation enforcement are generally hidden from the public eye, and regulatory agencies are not generally accountable as such. However, since different site-specific mining contexts often require individual permit approval, there is scope for bribery, which is endemic in the bureaucracies and industry of many developing countries. Even though there is a theoretical threat of mine closures due to noncompliance, most foreign mining firms know that their developing country host can least afford to lose the foreign exchange earnings from their activities. Therefore, the risk of closure due to environmental noncompliance of this type is considered relatively low.

Finally, environmental regulations often emerge that are contradicted by other economic and industrial policies. For example, several countries with tropical forests have recently introduced policies aimed at their conservation. At the same time, countries such as Brazil, Ecuador, and Colombia have parallel economic policies to promote industrial investment, especially by foreign firms, in remote areas of these countries. One example of this occurred when the government of Ecuador authorised RTZ Corporation to invest in a mine in one of Ecuador's national parks; RTZ later withdrew to avoid controversy over the issue. Similarly, in Brazil, where forest conservation policies (in part conditional upon European Community and World Bank loans) were in place, the Carajas smelters were fuelled by large amounts of charcoal from neighbouring forests.

In summary, command-and-control regulation tends to identify and deal with the symptoms (pollution) of environmental mismanagement rather than the causes (production inefficiency, human resource constraints, lack of technology and capital). It is also add-on and incremental in nature. Therefore, there is a tendency for it to emphasise end-of-pipe, add-on, and capital-intensive solutions (e.g., smelter scrubbers, mine water treatment plants, dust precipitators, etc.) for existing technology and work practices, rather than promote alternative environmental management systems or technological innovation. Regulation may also, to a certain extent, presuppose a static technology. If regulation is incremental, technical change may be incremental, involving the addition of numerous new controls at relatively greater cost and with more overall resultant degradation than if a new more radical change had been introduced in the first place. In turn, this regulatory approach may get a more uncooperative response from industry, which sees the rules always changing and their cost implications increasing. Furthermore, such regulation ignores the human resource elements of sound environmental management by emphasising specific pollution control technology rather than training, managerial approaches, and information diffusion.

Technical Change and Corporate Environmental Trajectories

Enterprises respond to environmental pressures in ways that have been characteristically slow and reflect the regulatory regimes and public climate of either their home country or foreign countries of operation. Their response has also depended on the nature of their operations in terms of, first, the mineral involved; second, the level of integration of mining and processing activities; third, the stage in the investment and operations cycle which its mineral projects have reached; and fourth, the internal economic and technological dynamism of the firm, that is, whether it has the financial, technical, and managerial capabilities to be an innovator.

After a period of using rather static technology, the mining and mineral processing industry is going through a phase of technical change, as dynamic firms are innovating by developing new smelting and leaching technologies to escape economic as well as environmental constraints. Rapidly evolving environmental regulatory frameworks in the industrialised countries and the prospects of their application, reinforced by credit conditionality, in the developing countries are stimulating this trend. Changed technological and environmental behaviour in this context is evident particularly in the large North American and Australian mining firms, but it is becoming increasingly apparent in firms operating in developing countries, for example, Chile, Brazil, and Ghana. However, it seems to be the new operators and dynamic private firms that are changing their environmental behaviour, while state-owned enterprises and small-scale mining groups in developing countries continue with some exceptions to face constraints regarding their capacity to change environmentally damaging practices.

It is somewhat inevitable that only those firms that are dynamic and have new project development plans are in a position to invest in the R&D required to develop environmentally sound alternatives or to raise the capital to acquire them from technology suppliers. Nonetheless, after a long period of only conservative and incremental technical change, alternative process routes for mineral production are being developed, and they are emerging as more economically efficient as well as less environmentally hazardous. Furthermore, firms are beginning to sell their technologies, preferring to commercialise their innovations to recoup their R&D costs rather than sell obsolete technology and risk shareholders' scorn or retroactive penalties as environmental regulations are increasingly enforced by the developing countries. Some of those firms have pushed technology even beyond the bounds of existing regulations and as a consequence are seeking to *increase regulation*—particularly on a worldwide scale—because they can meet the requirements and use their new environmentally sound technologies to their competitive advantage.

There is evidence that improving the environmental management of a mining operation may not necessarily be detrimental to economic performance and in some cases may even be of economic benefit. Furthermore, because environmental regulation is here to stay and bound to become more widely adopted, more stringent, and better enforced, the winners in the division of shares in the metal markets will not be those firms that avoid environmental control (only later to be forced to internalise the high cost of having done so), but will be those firms that were ahead of the game, those that played a role in changing the industry's production parameters, and those that used their innovative capabilities to their competitive advantage.

Environmental trajectories describe the evolutionary development of a firm's competitiveness and environmental compliance in response to different environmental and market pressures. The different types of environmental trajectory that mining firms might take are categorised in Figure 2. This diagram could be a planning tool for both firms and governments. It can help to evaluate the environmental and economic implications of applying different policies on corporate development.

Figure 2 argues that, contrary to the popular belief that a trade-off exists between production costs per unit of output and environmental damages (which presupposes a static technology), new technology is being developed that lowers environmental and economic costs and pushes forward the frontier of leading technology. Figure 2 is a simplified picture of different firm behaviour patterns towards the internalisation of the environmental costs (traditionally suffered by the pollutee or absorbed by the state) resulting from their mining activities. Firms in developing countries are placed in a generalised group behind the technological frontier (obviously there are several exceptions to this generalisation), and their behaviour is associated with higher environmental damages and low to high production costs (depending on the efficiency of their operations). The technological frontier is a generalised band within which most mining firms operate, to a greater or lesser extent absorbing the environmental costs resulting from their activity. Firms with ongoing operations, high sunken costs in existing facilities, and less dynamism in their technological behaviour would tend to experience environmental regulatory pressures as a cost burden, which would push them along a trajectory B^1–B^2–B^3 as they respond to successive regulations incrementally with generally expensive add-on controls. Other firms are innovating by building into the new generation of technology that they are developing both lower

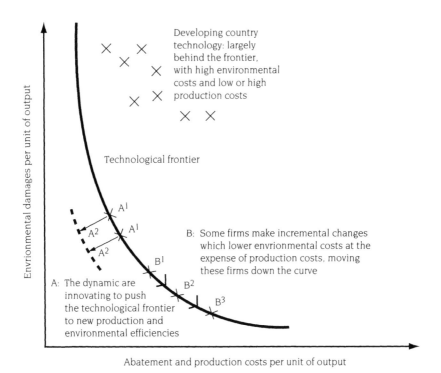

Figure 2. Environmental trajectories: a company's position in relation to the technology frontier will be affected by its current technology and its approach to innovation.

Source: Warhurst 1992a

production costs and less environmental damage (protecting themselves in the process from having to undertake costly add-on incremental technical change at later stages of their operation). In doing so they are pushing forward the technological frontier along a trajectory A^1–A^2.

However, market pressures—mainly a real decline in metal prices—combined with their economic inefficiencies, mean some of these firms are going bankrupt (a trajectory towards B^3). They will leave a legacy of environmental pollution behind, and as in the cases of, for example, COMIBOL (in Bolivia) and Carnon (in the United Kingdom), the burden of cleanup will fall on the state and society. Those firms that respond by innovating, moving towards A^2, are building into new generations of technology both improved economic and environmental efficiencies (protecting themselves in the process from having to undertake relatively more costly add-on, incremental technical change and rehabilitation at later stages in their operation). Indeed, freed from the incumbent costs of retrofitting sunken investments, greenfield plants in particular display new levels of dynamism— the latest best-practice technology incorporates both improved economic and environmental efficiencies.

Market and regulatory pressures suggest that the average firm will only survive in the new regime if it innovates. Even previously dynamic firms will need to keep their environmental trajectories moving ahead. Moreover, this implies a serious constraint on the regulatory process for two main reasons, which indeed distinguish mining firms from their manufacturing counterparts. First, an implied close-down due to regulatory burden does not signal the end of environmental degradation. Pollution in metals production is not all end-of-pipe. Rather it heralds a new era, since decommissioning, cleanup, and rehabilitation all pose significant environmental costs. Second, in very few countries are bankrupt operators liable for the cleanup of their 'sins of the past'. The United States with its Superfund liability laws is an exception. (See the Tilton lecture in this volume for a critical assessment of U.S. Superfund policy.) The policy challenge of the environmental imperative is therefore how to keep firms sufficiently dynamic to be able to afford to clean up their pollution and generate economic wealth.

Cleaning Up 'Sins of the Past' and Some Environmental Effects of Current Regulation

It is inevitable that those firms with long mining histories and extensive sunken investments in conventional mining and smelting facilities face the greatest technical, and therefore economic, challenges in cleaning up their past facilities and reducing pollution from their ongoing operations. For example, some firms in the central and southwest United States have found that dumps from former lead and copper mining operations have now created such serious acid mine drainage and toxic seepage that the government has placed them on its Superfund list, which obliges multimillion dollar sums to be spent on their cleanup. The government then targets previous owners of the mine, often the richest, making them liable. If a firm has already closed down a mine and written off the investment, and perhaps is struggling in the current economic climate to manage a new project, it is clear that the costs of such a Superfund indictment, and the legal costs involved in answering it, can be quite crippling.

For example, the Smuggler Mountain lead-mining site in Colorado has a serious acid mine drainage and toxic seepage problem (EPA 1989). Its old lead and cadmium mine workings have apparently contaminated soils and groundwater in neighbouring residential areas, requiring a major cleanup operation, the secure repositioning of the toxic waste, and the establishment of monitoring mechanisms and pollution controls to prohibit further contamination. The estimated cost of this project by EPA is currently US$4.2 million.[3]

Although there may be economic opportunities associated with such cleanup operations, such as the recovery of extra metal values from acid mine drainage, the commercialisation of innovative water treatment methods, or the innovative use of tailings material, these may not be recouped by the mining firm itself. Furthermore, firms that previously have had no links to the facility may be nervous of getting involved in case any liability may be passed on to them.

This is one reason why some mining firms, which need to clean up their past operations, object to such retroactive regulation and suggest that such restrictions and control threaten their existence. An interesting and illustrative case concerns the respective responsibilities of government and industry for the management of mine closures and the rehabilitation of old tin, copper, and silver mining and smelting opera-

---

[3]Another Superfund-listed mine site is the Silver Bow Creek Site in Butte Area, Montana, which for over 100 years has been mined for silver, copper, gold, and zinc, resulting in severe water and soil contamination and the disruption of local ground and surface drainage water patterns. Currently groundwater is flooding the mine, becoming highly acidified in the process, and absorbing high concentrations of iron, manganese, arsenic, lead, cadmium, copper, zinc, and sulphate. This toxic seepage is threatening Silver Bow Creek, a major river in the region. The cleanup and remedial action is extensive and involves detailed diagnostic analysis and monitoring, water and tailings containment, water treatment, and soil treatment. Major mining companies and individuals, all past owners, are implicated, including: Atlantic Richfield, AR Montana Corporation, and ASARCO. The cleanup cost is estimated to be in the region of several million dollars, to be confirmed once the precise plan of remedial action is determined (EPA 1990).

Another Superfund-listed old mining site is Gregory Tailings in Colorado (EPA 1986). It was a gold mine exploited during the late nineteenth century. Waste was placed in inadequate tailings dams, and resultant leakages have contaminated local water supplies and soils with acidic waters containing copper, zinc, nickel, cadmium, and arsenic. The cost of the cleanup and water treatment has yet to be confirmed, but the strengthening of the tailings dam is estimated at over half a million dollars.

tions in Cornwall, United Kingdom. The observations that mine closure and rehabilitation were proceeding very inefficiently in the absence of an adequate regulatory framework have been borne out by the recent flooding of polluted acid mine water from the Wheal Jane mine near Truro, Cornwall (Warhurst 1992). This highly acidic cocktail of dissolved metals, including copper, lead, cadmium, tin, and arsenic, entered the Carnon and Fal rivers at a rate of two to four million gallons a day and has spread throughout the surrounding estuary and coastal areas. The tourist industry and fisheries have been threatened and local well-water supplies destroyed.

Many have argued that legal responsibility for the mine discharge rests with the current owners, Carnon Consolidated (a management buy-out in 1990 from RTZ Corporation, which had bought it seven years previously from Consolidated Goldfields). However, the National Rivers Authority (NRA), the relevant U.K. regulatory body, publicly admitted that they were aware six months previously that this disaster could happen but were unable to prevent it, since their policy remit did not cover preventative action. The disaster occurred following the withdrawal of a government grant to the Wheal Jane mine, which has always required pumping to prevent it from flooding. The withdrawal of funds meant that plans had to be shelved for turning the site into a golf course and leisure centre that would help fund the pumping and maintenance costs at the mine, which were £100,000 per month. The consequent lack of finance forced Carnon Consolidated to turn off all the pumps on January 4, 1992. Since they had officially abandoned the mine, they denied responsibility for any ensuing flooding and pollution, a claim complicated by the fact that numerous underground shafts from other abandoned mines in the region also provided conduits for the polluted floodwater. The NRA has stated its intention to prosecute Carnon, but loopholes in the recent 1991 Water Resources Act mean there is no U.K. regulation to ensure that previous mine owners bear the financial liability for cleanup measures (unlike the Superfund legislation in the United States). Abandoned mines are specifically exempted from cleanup liability. Furthermore, the NRA itself has no remit or budget to treat pollution, particularly on this scale.

The NRA has opened an old dam to hold and treat the contaminated water. It is also analysing the potential of biotechnology to assist in the cleanup. For example, certain microbes bred in a slurry of cattle excrement have been shown to be metal-absorbing. Similarly, the creation of wetlands containing plants whose roots absorb metals is another possible long-term solution. However, such solutions are based on piecemeal research being undertaken in this area in different research institutions (e.g., the Colorado School of Mines in the United States, the Canadian Centre for Mineral and Energy Technologies in Canada, the Centro de Tecnologia Mineral in Brazil), and none of these techniques has yet been proven commercially. The NRA is approaching the government to help pay for the cleanup, which may take decades and could cost over £1 million. The ultimate financial responsibility for mine cleanup in the United Kingdom may therefore lie with the taxpayer.

Environmentalists in the United Kingdom are pressing for the law to be changed so that firms that abandon mines are liable for any resulting pollution. However, it has been suggested that such a change would make life difficult for British Coal. The NRA has already analysed pollution from rising waters in old coal mine workings, which is affecting rivers in South Wales and Yorkshire. The NRA has established that the capital costs of treating the ten worst cases in Yorkshire are estimated at more than £10 million. Indeed, as long ago as 1981, the Royal Commission on coal and the environment (which prompted the Flowers Report) recommended that the costs of remedial action for existing mines abandoned by the National Coal Board be met by central government. However, the tightening of U.K. laws to ensure that the bills for such pollution are paid by previous owners could have serious implications for the current government's plans to privatise the coal industry.

Similar pollution liability issues are also being faced by governments in developing countries currently engaged in privatising their state mines—Bolivia, Peru, and Chile are cases in point. In Peru, for example, the legacy of past pollution, particularly from toxic tailings along the river below Centromin's La Oroya and Cerro Pasco facilities, was preventing the government from selling those enterprises, since the cost of cleanup rendered the investments uneconomic and unattractive to foreign capital. It was therefore agreed that the investment contract for buying these operations would protect the foreign partner from liability for previous environmental damage. They would start with a clean slate, as it were, with generous lag times regarding the introduction of new environmental controls to reduce ongoing pollution. This means that the economic burden for cleanup falls on the state but, in developing countries, that means on society. Where capital is scarce, cleaning up pollution problems from past decades affecting remote rural communities is of low priority.

This poses a policy dilemma. If the government does not waive liability for past pollution, the privatisation schemes, vital to the future of the economy, will not succeed, since foreign partners would not be interested relative to other available investment opportunities in mining. Moreover, international firms are particularly wary of falling prey to new retroactive liability laws and punitive tariffs in their import markets, particularly given the enormity of the cleanup involved. The responsibility for cleanup therefore lies formally with the state. Should loan-conditionality put pressure on the government to clean up using precious capital resources? Should cleanup funds be established and incentives provided to prompt local industry to develop technical solutions? Should aid programmes provide technical assistance and training in cleanup? Should new investors be taxed for old pollution? Clearly the optimal outcome would be for government and donors to provide incentive programmes for local firms to seize the commercial opportunities available. Nuñez (1992) clearly documents the extensive range of local capabilities which could be harnessed.

While Superfund in the United States may theoretically be successful, given that one is targeting local investors and traceable firms, it may be more difficult to target and litigate against previous mine owners (prior to nationalisation) in developing countries. Searching out the foreign investors responsible for the many old and abandoned mines may be difficult since most have long since returned home, and local miners have limited resources, which makes the task of determining liability and enforcing cleanup a daunting one.

Firms that are being forced through environmental pressures to deal with pollution problems in their existing operations have been observed to react both defensively and in innovative fashion, depending on the challenges posed, their economic well-being, and internal dynamism. For example, depending on the level of enforcement of the regulatory regime, some mining firms, particularly those operating in developing countries, may prefer to pay financial penalties and fines for affected water and air quality. These may amount to less than the cost and effort involved in remedial action, such as water treatment, and considerably less than the costs involved in innovation or the incorporation of pollution controls. In some instances, as discussed above, the state pays those remedial costs, and it is clear that the state may be subsidising the profit of foreign mining firms at the expense of environmental degradation. Sometimes that trade-off is influenced by the state's absolute dependence on the foreign firm as a source of foreign exchange and government revenue.

Indeed, a number of mining firms perceive that environmental regulation imposes a cost burden on their operations that threatens their profitability. They may then enter into negotiations with the state to arrive at a 'stay of execution' or to devise a plan for implementing controls. However, as regulation becomes increasingly strict, backed up by more sophisticated monitoring devices and data processing, firms are being pushed to take remedial action in both the industrialised and developing countries. Data from the United States suggest trends that may be followed elsewhere. According to Coppel (1992), sulphur dioxide emission controls have resulted in 'substantial capital expenditure' for U.S. copper smelters and increased operating costs due to add-on acid plants. Present levels of environmental control entail capital and operating costs of between ten and fifteen cents per pound of copper. However, the United States has lost substantial smelting capacity. It has been reported that eight out of sixteen smelters operating in the United States in the late 1970s have closed permanently, 'most because the capital investment to meet regulations was unwarranted given current and anticipated market conditions' (Office of Technology Assessment 1988, 16).

Indeed, the evidence from studies in the United States shows that environmental compliance does not distort significantly the economics of new mineral projects, but does place a considerable cost burden on ongoing facilities for either retrofitting or cleanup on mine and plant closure. The U.S. Bureau of Mines (1990, 28) has estimated direct environmental operating costs for smelting facilities with emission controls; retrofit capital costs were estimated to be of the order of US$150 million per facility, or 5.6 cents per pound of copper produced. According to Coppel 1992, the overall cost penalty, including capital invested, to the producer for implementing the new smelter and sulphur dioxide capture facilities was estimated to be 7.5 cents per pound after deductions of a 1.3 cents per pound of acid credit. The operating costs for individual smelters ranged from 10 to 15 cents per pound of copper, and the average operating cost in 1987 was 12.3 cents per pound. Of this amount, 26 percent, or 3.2 cents, was calculated by the U.S. Bureau of Mines to be the cost burden of compliance with environmental, health and safety regulations.

Dynamic Innovators: Technical Change to Improve Environmental Management

Although some mining firms resist the application of environmental regulation to their existing operations, a growing number of dynamic innovative firms are making new investments in environmental management because they see an evolution toward stricter environmental regulation. Free of the encumbrance of sunken investments in pollutant-producing, obsolete technology or with significant resources for R&D and technology acquisition, they have chosen either to develop more environmentally sound alternatives or to select new, improved technologies from mining equipment suppliers, who are themselves busy innovating. Increasingly, these new investment projects are incorporating both improved economic and environmental efficiencies into their new production processes, not just in terms of new plants or items of technology, but also through the use of improved environmental management practices. Some examples of these are discussed below in three categories: smelter emissions, gold extraction, and waste management.

Smelter emissions. *Inco Ltd. (Canada).* At one time one of the world's highest-cost nickel producers, Inco was until recently the greatest single point source of environmental pollution in North America. This was due to its aged and inefficient reverberatory furnace smelter technology, which spewed out excessive tonnages of sulphur dioxide emissions. Inco Ltd. had reached the limit of improving the efficiency of this obsolete technology through incremental technical change at the same time as the Ontario Ministry of the Environment began an intensive sulphur dioxide abatement programme to control acid rain. These factors prompted Inco to invest over Can$3,000 million in a massive R&D and technological innovation programme (Aitken 1990).

Under the Canadian acid-rain control programme, Inco is required to reduce sulphur dioxide emissions from its Sudbury smelter complex from the current level of 685,000 tons per year to 265,000 tons per year by 1994, a 60 percent reduction. To achieve this, Inco plans to spend Can$69 million to modernise milling and concentrating operations and Can$425 million for smelter sulphur dioxide abatement. The modernisation process will include replacement of its reverberatory furnaces with a new, innovative, oxygen-flash smelter, a new sulphuric acid recovery plant, and an additional oxygen plant. By incorporating two of these flash smelters, the firm plans to reduce emissions by over 100,000 tons per year in 1992 and by 1994 to achieve the government target levels of 175,000 tons per year. Other environmental benefits include a cleaner, safer work environment (*Mining Journal* 1990).

Inco is now one of the world's lowest-cost nickel producers, and, like other dynamic firms that are responding to environmental regulation through innovation, Inco is seeking to recoup R&D costs through an aggressive licensing effort in other copper and nickel processing countries. More than 12 percent of Inco's capital spending during the last ten years has been for environmental concerns (Coppel 1992).

Kennecott Corporation (Utah, United States). A new smelter project has recently been launched by Kennecott (a subsidiary of RTZ Corporation) with the dual aims of setting a new standard for the cleanest smelter worldwide and achieving improved cost efficiencies in processing its ore. Advantages include the capture of 99.9 percent of sulphur off-gases (previous levels were 93 percent). Sulphur dioxide emissions will be reduced to a new world best-practice level of approximately 200 pounds per hour, less than one twentieth of the 4,600 pounds per hour permissible under Utah's current clean-air plan. The investment is US$880 million, resulting in 3,300 new construction jobs and the investment of US$480 million in local firms through project development contracts (Kennecott 1992).

The proposed Garfield smelter will expand the concentrate processing capability to the level of mine output (about one million tonnes of copper concentrate per annum) at more than half the previous operating costs. It represents the first time oxygen-flash technology will be applied to the conversion of copper matte to blister. According to Kennecott 1992, the two-step copper smelting process consists of smelting furnaces that separate the copper from iron and other impurities in a molten bath, followed by converting furnaces where sulphur is removed from the molten copper. A new technology known as flash converting will then be utilised in the second step of the process at the new smelter. This unique technology was developed by Kennecott in cooperation with Outokumpu Oy, a Finnish firm and a leader in the supply of smelting technology. Essentially the new technology eliminates the open-air transfer of molten metal and substitutes a totally enclosed process for producing molten metal. Flash converting has two basic effects: first, it allows for a larger capture of gases than the current open-air process; second, it allows the smelter's primary pollution control device—the acid plant—to operate more efficiently. The smelter will include double-contact acid plant technology.

There will be other environmental benefits from the new smelter as well. Water usage will be reduced by a factor of four through an extensive recycling plan. Pollution prevention, workplace safety and hygiene, and waste minimisation will be incorporated into all aspects of the design. In addition, the smelter will generate 85 percent of its own electricity by recovering energy as steam from the furnace gases and emission control equipment. This eliminates the need to burn additional fossil fuel to provide power. The new facility will require only 25 percent of the electrical power and natural gas now used per tonne of copper produced.

The copper refinery's planned modernisation and expansion modifications include major electrical system changes, material handling system improvements, and new electro-refining cells. In addition, a new state-of-the-art precious metals refinery will be built. The refinery will be able to process the entire output of the new smelter.

Gold extraction. *Homestake's McLaughlin Gold Mine (California, United States).* The McLaughlin gold mine, opened by Homestake Gold Mining Company in 1988, is perhaps the best example of a new mine and processing facility that has been designed, constructed, and operated from the outset within the bounds of probably the world's strictest environmental regime. Environmental efficiency is built into every aspect of the gold mining process, in terms of innovative process design criteria, fail-safe tailings and waste disposal systems, and extensive ongoing mine rehabilitation and environmental monitoring systems. The mining operation therefore combines innovative technologies with best practices in environmental management. The most interesting conclusions drawn by the author from site visits and discussions with the firm's environmental officers is that most of these environmental management initiatives have not resulted in any substantial extra cost, and indeed many of these procedures have apparently improved the efficiency of the mine, affecting positively the economics of the overall operation.

For example, before the mining operation began, an extensive environmental impact analysis and survey were undertaken. All plant and animal species were identified and relocated, ready for rehabilitation on the completion of mining operations. The survey also measured in detail prior air, soil, and water quality characteristics and flow patterns to provide the baseline for future monitoring programmes. Assaying was undertaken not just of the gold ore, but also of the different types of gangue material and waste, so that waste of different chemical compositions could be mined selectively and dumped in specific combinations to reduce its capacity to generate acid mine drainage. Local climate conditions were evaluated to determine the frequency of water spraying to reduce dust, and evaporation rates were evaluated to control the water content and flood risk potential of tailings ponds. The tailings ponds themselves are constructed on specially layered impermeable natural and artificial filters, with high banking to prevent overflow and with secondary impermeable collecting ponds in the rare case of flooding.

Unlike other mining projects, where rehabilitation is seen as a costly task to be undertaken at the end of a mining operation, often when cash flows are lowest as ore grades decline, at Homestake rehabilitation began immediately and is an ongoing activity. Not only does this serve to spread expenditure more evenly over the life of the mine, but it enables the more efficient utilisation of truck and earth-moving capacity, as well as of relevant construction personnel. This means that as soon as work piles have reached a certain predetermined dimension, soils (previously stripped from the mine area and stored) are laid down and revegetation is begun. Although mining has been under way for only three years, extensive areas of overburden and waste have already been successfully revegetated—immediately reducing environmental degradation and negative visual impacts. In addition to these built-in environmental control mechanisms, Homestake Mining Company has sophisticated environmental monitoring procedures in place. This means that seepages, emission irregularities, wildlife, and vegetation effects can be detected and rectified immediately, which in the long term reduces the risk of expensive shutdowns in operations, costly court cases (e.g., if water toxicity results), and the need for treatment technologies.

Waste treatment. In the minerals industry, considerable waste is produced in the form of overburden, marginal ore dumps, tailings, and slags. Much of the toxicity associated with that waste is a direct result of the loss of either expensive chemical reagents or metal values. Currently, public policy has not taken up the challenge to promote and direct R&D in the area of waste reduction and treatment innovations. One interesting area is the application of biotechnology to waste treatment (Warhurst 1991a).

Homestake's Mine at Lead (South Dakota, United States). In the area of water treatment, Homestake turned to its advantage regulatory pressure to clean up a cyanide seepage problem. Its own R&D staff developed a proprietary biological technique to treat the effluent, leading to the recovery of local fisheries and water quality in the mine's vicinity at Lead (Crouch 1990). Homestake is now actively commercialising the technology, which could be widely applied at other gold leaching plants.

Exxon's Los Bronces Mine (Chile). A mining project in Chile, Los Bronces, is to be expanded into one of the world's largest open-pit copper mines and consequently required the stripping of very large tonnages of overburden and low-grade ore. Before mine development, the Chilean government warned Exxon that it would be imposing financial penalties for the water treatment costs on account of expected acid mine drainage from the overburden of low-grade ore dumps into the Mantaro River, the source of Santiago's drinking water. This threat became the economic justification for a bacterial leaching project at the mine. Indeed, the feasibility of this bacterial leaching project was particularly illustrative of the profitability of leaching copper from waste at the same time as prohibiting otherwise naturally occurring pollution (acid mine drainage). Over a billion tons of waste and marginal ore below the 0.6 percent copper cutoff grade are expected to be dumped during the project's lifetime. The waste would have an average grade of 0.25 percent copper and would therefore contain a lucrative 2.5 million tons of metal worth approximately US$3,500 million at 1985 prices (Warhurst 1990). The study demonstrated that with a 25 percent recovery, high-quality cathode copper could be produced profitably, at 39 cents per pound, by recycling mine and dump drainage waters through the dumps over a twenty-year period. This was shown to have the double advantage of both extracting extra copper and avoiding government charges for water treatment. At the same time, both investment and operating costs were less than two-thirds of the estimated costs for a conventional water treatment plant, which would not have had the benefit of generating saleable copper. The Los Bronces mine thus demonstrates the potential economic benefits of building environmental controls into mine development.

In conclusion, these few examples suggest that dynamic firms are not closing down, reinvesting elsewhere, or exporting pollution to less restrictive developing countries; rather, they are adapting to environmental regulatory pressures with innovation and by improving and commercialising their environmental practices at home and abroad.

Policy Implications for Mineral-Producing Countries

Technical Change and the Environmental Trade-Off

Evidence suggests that, at least during the 1980s, environmental policies have not been a major factor in determining where a mining firm will target exploration and subsequent investment activities. Geological potential remains of primary importance, which is not to underestimate that in some cases the approval and permitting process is a major cost of compliance.[4] This would suggest that developing countries are not seen as pollution havens and that the industrialised countries' environmental regulations are not stifling new mining investment. Indeed, there are currently several new gold projects in the process of development in California, which has probably the world's strictest environmental regulatory regime. Although environmental policies may not negatively influence the investment activities of dynamic, adaptive mining firms, the latter still seek to play a role in determining the detail and focus of relevant legislation so that new regulatory frameworks also reflect, as far as possible, their corporate interests. During preliminary fieldwork by the author in North America and Europe, it became evident that this task was an important function of many of the firms' newly appointed vice-presidents for environment, directors, and environmental affairs representatives.

If one understands the new environmental pressures being placed on the mining industry in the industrialised countries in the context of hard-won survival following a prolonged period of low metal prices, which gave significant market advantages to their lower-cost competitors in the developing countries, then it is possible to understand the recent lobbying by some firms for industry-wide international environmental standards. Although international standards may not pose too much of a problem for the economics of new mining projects in the developing countries, our analysis here suggests there could be major costs incurred by any older, ongoing operations. Controlling the latter's pollution problems would in most cases require major water treatment

---

[4]Johnson (1990) ranked corporate criteria in selecting countries for exploration. See also Eggert 1992.

plants, strengthening and rebuilding tailings dams, add-on scrubbers and dust precipitators, etc. The imposition and strict regulation of international environmental standards could make some developing countries' mineral production uneconomic, thus swapping one social cost (environmental pollution) for another (unemployment and poverty, and indeed cleanup, given the absence of liability laws). This is not to dispute the need for improved environmental controls, particularly in the developing countries, but rather to show the complexity of the process that sets the environmental agenda.

Most planned mines (including existing mine expansions) and available reserves are located in the developing countries. Furthermore, after a period of mineral production monopolised by state-owned mining firms (with some exceptions), many developing countries are now embarking upon a phase of liberalisation and have legislated a number of laws and incentives to promote foreign investment. In many cases, those investments are being partly financed by credit that is conditional upon good environmental practice and prior environmental impact analysis. The upshot is that this trend in technical change may be to the benefit of the developing countries, in that it may enable them to reduce the trade-off between higher environmental costs and lower production costs. This may mean that, at least in the case of new mineral projects, there may be a wider range of more environmentally sound and economically efficient technologies available to them.

Indeed, new flexible-scale, lower-cost, less-hazardous hydrometallurgical (leaching) alternatives to conventional smelting may be further to the advantage of developing countries, improving value-added from their mineral production. For example, processing up to the stage of a final saleable metal product can be undertaken at the mine site—while in conventional process-routes a smelter would require feed from at least ten large mines for full capacity utilisation, and ore may have previously been exported to foreign smelters with consequent loss of by-products and entailing charges for the treatment of pollutant elements.

However, this new prospect of environmental security may have its own costs, which require careful analysis. For example, depending on how technology transfer agreements are drafted and managed, such new environmentally friendly investment may herald indebtedness, bankruptcy of local equipment suppliers and engineering firms, and the loss of employment, etc, reinforced by aid conditionality. On the other hand, smelting, concentration, and leaching innovations are being developed by international firms, such as Outokumpu, Mitsu-bishi, Kennecott, Inco, Cyprus Minerals, and Homestake, which are adapted to new and prospective regulations in the industrialised coun-tries. These trends may oblige developing countries for both economic and environmental reasons to export only semiprocessed minerals or raw materials, reinforced by credit conditionality, new international regulatory agreements, and trade tariffs imposed on imports of metals produced not using a predetermined best available technology.

Technology Policy for Environmental Management

Environmental behaviour correlates most closely with a firm's capacity to innovate, rather than its size, origin, scale and scope of operations, or ownership structure. For example, government policy over time has resulted in a failure by state enterprises to reinvest capital in human resource development, repairs and preventative maintenance, R&D, and technology development. Managers became bureaucrats rather than entrepreneurs (Jordan and Warhurst 1992). This factor, combined with cumulative inefficiencies, a poor waste management strategy lead-ing to metal and reagent losses, and scarce resources, means that envi-ronmental mismanagement is endemic. It is a structural problem and one not readily solved by recourse to regulation, punitive tariffs, or even the simple act of purchasing an environmental control technology.

The cases of Comibol in Bolivia and Mineroperu and Centromin in Peru bear witness to this, as does the case of private firms, such as Carnon Consolidated in the United Kingdom. However, Codelco and Enami, the state enterprises of Chile, have invested in developing their innovative capabilities both within the industry and through historically close links with local R&D institutions and universities. Although new regulations currently pose a significant technological challenge to these firms, efforts are being made to develop the required human resources and to implement substantial technical change. For example, Codelco is now at the forefront of metals biotechnology and has made consider-able investments in new solvent-extraction/electrowinning technology. In addition, Enami is planning to replace its reverberatory furnaces with modern flash-smelting technology at an estimated cost of US$300 mil-lion, 'largely motivated by the need for environmental improvements' (U.S. Bureau of Mines 1991; also Coppel 1992).

Environmental degradation from garimpo-type operations (that is, small-scale and largely illegal) is also related to the miners' incapacity to innovate through a lack of access to capital, technology, skills, and information. Scale further complicates the choice of optimal low-waste, high-metal-recovery technology. With few exceptions, however, the private sector has so far shown itself to be most innovative and therefore most capable of improving environmental management. In several cases, improved environmental management would have been brought about irrespective of regulation due to market pressures to introduce new, more efficient, low-waste technical change.

This is not to argue against regulation but rather to recommend a more sophisticated public policy approach through first, the definition of regulatory goals—something to aim at—and second, an informed technology policy to guide and stimulate those firms along the fastest, most efficient route to achieving those goals. These policy mechanisms are discussed in detail below.

Policy to Promote Environmental Innovation. There are two types of policy mechanisms that can be used to promote environmental innovation in industry. The first set includes expenditure programmes to support R&D, environmental engineering, clean technology development, and training in environmental management. The second is a set of incentives to stimulate and reward firms for environmental innovation.

Supporting Clean Technology Development. Technology policy mechanisms to assist the funding of clean technology development include the targeting of R&D in selected areas of pollution prevention. It would include the cofunding of R&D projects involving interindustry and industry-university/research institution collaboration. Such programmes could be supported through easily accessible, centrally compiled, information dissemination programmes about moving technological and regulatory frontiers.

Targeting firms as innovators of pollution prevention technology requires a number of important changes in policy thinking. A multimedia approach is required, since pollution prevention requires changes in process technology, not the addition of off-the-shelf, end-of-pipe controls, which tend to shift pollution from one media to another. This requires a range of engineering skills harnessed to deal with the reduction or elimination of the pollutant at source (independently of where it may ultimately be discarded). New technology must be designed to deal not only with water and air quality and waste, but also with workers' health and consumer products' safety. This means that training for research and development engineers *in* industry should be a critical element of pollution prevention policy.

Technology hardware is only one part of the equation. Of equal importance is the need for organisational change. Much can be learnt from the manufacturing sector regarding the development and success of 'Lean Production' and related Japanese work methods, such as 'just-in-time' inventory control, waste reduction throughout the system, total quality management, and statistical process control. Lean Production is driven by a simple principle: the elimination of all costs incurred that do not add competitive value to a product. Secondary principles include the reduction of waste, utilisation of space, the elimination of inventories, and the integration of quality control within the production process. The implementation of Lean Production characteristically results in the reduction of managerial roles, with increased responsibility being given to engineers and workers, and a concomitant increase of multitask activities (Womack, Jones, and Roos 1990). The implications of applying Lean Production principles to the mining industry, or of developing radical process innovations with similar effects, would be remarkable. A combination of markedly lower investment and production costs, combined with the halving of mine development times and mine life, could have significant implications for the competitive structure of the industry as well as reducing associated negative environmental and social effects. Few mining firms have taken these ideas on board. Those that have considered alternative organisational methods include CRA (Australia), Homestake's McLaughlin Mine (California, United States), and Scuddles Mine of the Poseidon Group (Australia).[5]

Firm Incentives for Environmental Innovation. Changes may be necessary in taxation policy in order to promote environmental innovation. According to Ashford 1991, the United States currently gives taxation incentives in the form of accelerated depreciation for pollution control equipment, thus supporting end-of-pipe pollution control. However, investments in new production technology are not similarly treated, so that dollar-for-dollar a firm would be better off buying from an environmental technology vendor than developing process changes. Direct taxation incentives could relate to: investment in technological or

---

[5]Scuddles has implemented an innovative, multiskilled approach to human resource development at its underground mine in Western Australia (*Mining Magazine* 1991).

organisational change; R&D; engineering projects and training in specific areas of environmental management; and the posting of bonds up front for future pollution prevention or reclamation on closure. Punitive taxation on reagent use or energy use requires the careful consideration of its effects on both competitiveness and firm behaviour, since different deposits, due to geology and chemistry, pose quite different implications for energy and reagent consumption patterns. Taxation in this context of ongoing operations may be perceived by operators as prejudicial and unfair.

This suggests that flexible taxation provisions that allow and encourage innovative responses by industry are needed to complement strict standards and regulatory goals. Innovative firms should be able to use environmental regulations to their competitive advantage. Benefits to such firms arise from the tightening of 'technology-forcing' regulation so that other firms are stimulated either to invest in new technology or license (or purchase) the innovator's technology (thus enabling the innovator to recoup some of the costs of its initial investment in R&D). Regulatory authorities need to be seen to respond in this way. Moreover the rate of technological advance in pollution control is probably, at least for the informed regulator, the most useful criterion on which to judge the effect of environmental policies.[6] Training for regulators, including industrial experience and salary packages commensurate with corporate counterparts, will also be important. As shown in Figure 2, the pushing forward of the technological frontier in this way has the effect of increasing the economic and environmental competitiveness of innovators. Consequently, the market conditions governing metals production also can change to the innovator's advantage.

Milliman and Prince (1989) analyse five regulatory approaches: direct controls, emission subsidies, emission taxes, free marketable permits, and auctioned marketable permits. They found that on a relative basis, direct controls—which are the most common regulatory tool—provide the lowest incentives to promote technological innovation in firms. Free permits and emission subsidies were also found to provide low incentives, while emission taxes and auctioned permits provided the highest incentives, by virtue of rewarding the innovator with positive gains beyond the firm's own application of the technology

---

[6]This view is reinforced by a growing number of researchers including Milliman and Prince 1989; Kneese and Schultze 1978; and Orr 1976.

through the benefits accrued from its diffusion to other firms.[7] This is not surprising since for polluters with high costs of abatement, it will be cheaper to buy permits than to reduce their emissions; polluters with low abatement costs will sell permits accordingly. Firms therefore have a constant incentive to cut emissions, since this allows them to sell permits. As such, tradeable permits have the advantage over pollution charges that they can guarantee the achievement of particular pollution targets, since the authorities control the number of available permits.

Finally, incentives need to be found to stimulate auxiliary firms to develop and commercialise innovative cleanup technologies, including remining techniques. In developing countries particularly, the market for such activities is vast, and donor agencies and development assistance grants could play a key role in stimulating such investment.[8] In the United States particularly, liability regulation will need to be reassessed to remove the current barriers to remining and treating existing mining waste.

In summary, regulators need to assess the rewards to the innovator of technology diffusion within the sector, as well as of innovation applied in the firm. This is the subject of the following section.

The Role of Technology Transfer in Environmental Management. Technological and managerial capabilities are required not only to innovate or to deal with new and emerging technologies, they are also vital to an environmental management strategy using existing technology due to the need to resolve pervasive inefficiencies. Technology transfer and technology partnership through joint-venture arrangements or strategic alliances are one way to build up technological and managerial capabilities to overcome these constraints. This is particularly true in the developing country context, although such strategic alliances are emerging in all the major mineral-producing countries. Recent examples of collaborative partnerships in environmental innovation include: Outokumpu and

---

[7]The arguments explaining these conclusions and the qualifications to be taken into account are long and complex and for brevity's sake are not repeated here. The reader is therefore referred to the original text, Milliman and Prince 1989.

[8]For example, over two-thirds of the current mineral reserves of Bolivia are in dumps and tailings (Jordan and Warhurst 1992). Furthermore, in many developing countries, such as Peru, there are many small- and medium-scale dynamic firms that supply a range of inputs to the mineral sector and could, with incentives, expand their activities to the environmental arena (Nuñez 1992).

Kennecott; Outokumpu and Codelco; Cyprus Minerals and Mitsubishi; Comalco, Marubeni Corporation of Japan, and the Chilean power firm Endesa; Battle Mountain of United States and Inti Raymi of Bolivia; and Comsur of Bolivia and Buenaventura of Peru.

However, there is a need to broaden the common concept of technology transfer to achieve the desired result of a real transfer of environmental management capability. Traditionally, technology transfer has meant a transfer of capital goods, engineering services, and equipment designs—the physical items of the investment, accompanied by training in operating equipment. As a consequence, the innovative capacity of recipients is undeveloped, and they remain purchasers and operators of imported plant and equipment. This is especially the case in developing countries, as recipients become dependent upon their suppliers to make changes or improvements to successive vintages of technology. Contractual conditions may reinforce this situation. New forms of technology transfer in environmental management need to go further to embrace, first, the knowledge, expertise, and experience required to manage technical change of both an incremental and radical nature and, second, the development of human resources to implement organisational changes to improve overall production efficiency, energy efficiency, and environmental management throughout the plant and facility, from mine development through production to waste treatment and disposal.

This new concept of technology transfer includes innovative approaches to training and skill acquisition within industrial enterprises in the areas of environmental R&D, engineering, management, on-the-job training, trouble-shooting, repair and maintenance, and environmental auditing.

In global industries such as mining, international firms contribute significant amounts of managerial and engineering expertise through joint ventures and other collaborative arrangements. Empirical research on other sectors, however, demonstrates that there exists considerable potential to increase those contributions without adversely affecting the supplier's strategic control over its proprietary technology (Bell 1990; Warhurst 1991a, 1991b; Auty and Warhurst 1993). Such an approach was at the heart of the strategy of China's National Offshore Oil Corporation, which targeted specific major oil firms and required them, under technology transfer agreements within investment contracts, to transfer the capabilities to master selected areas of technology (Warhurst 1991b). Another interesting example is the Zimbabwe Technical Management

Training Trust. It was founded by RTZ Corporation in 1982 with the aim of training professionals from the South African Development Community in technical management and leadership. It effectively combines academic and on-the-job training in both home and overseas operations, providing possibilities for accelerated managerial learning by being exposed to problem-solving situations with experienced colleagues in a range of challenging technical scenarios.

It would be quite feasible to build similar in-depth training programmes, concentrating on human resource development in environmental management, into many of the proposed and prospective mineral investment projects throughout the world. It cannot be overemphasised that all technology transfer and training efforts incur a set of costs for the supplier, and these must be covered to ensure optimal results. The danger of not budgeting for these costs would be to resort to a training programme in operational skills rather than in technology mastery skills. Corporate partners, the government, and, in the case of developing countries, donor agencies or development banks could assist in the financing of these schemes.

There already exists a range of commercial channels through which mine operators can purchase capital goods, engineering services, and design specifications; however, the market for knowledge and expertise, including training programmes, is less mature. It is the active development of this market that will reward innovators of pollution prevention technology. Bilateral and multilateral agencies, development banks, and government organisations can play a major role in improving this situation. Agenda 21, one of the main outputs of the United Nations Conference on Environment and Development, proposes two programmes of relevance (Johnson 1993; Skea 1994), which should also lead to greater involvement by industry. The first programme encourages interfirm cooperation with government support to transfer technologies that generate less waste and increase recycling. The second programme on responsible entrepreneurship encourages self-regulation, environmental R&D, worldwide corporate standards, and partnership schemes to improve access to clean technology.

The preceding analysis suggests there is the need for a two-tier policy approach. There is one set of challenges for the ongoing minerals industry, which must encompass the findings above regarding production inefficiency and its environmental consequences and the cleanup requirements for mine shutdown and plant decommissioning. Another set of challenges concerns the policy needed to build environmental management and the flexibility to engage in further environmental inno-vation into investment and expansion projects from the outset.

Conclusion

The public policy challenge is, therefore, how to keep firms sufficiently dynamic to be able to afford to both clean up their pollution and gen-erate economic wealth through innovation and sustainable-environ-ment management practices. The achievement of improved production efficiency and environmental management, particularly in developing countries, will in turn be dependent upon the extent to which far-reach-ing technology transfer and training clauses are built into the joint ven-tures and new investment arrangements which characterise the industry and whether banks, donor organisations, and governments demon-strate responsibility by providing the appropriate lines of credit and technical assistance in support of such objectives. Environmental regu-lation would be one element of that policy and would provide the goal posts for site-specific best practice in environmental management. Technology policy to promote technical change through technology transfer and human resource development would lie at its heart.

References

Aitken, R. 1990. Personal communication.

Ashford, N. A. 1991. Legislative Approaches for Encouraging Clean Technology. *Technology and Industrial Health.* 7 (516): 335–45.

Auty, R., and A. Warhurst. 1993. Sustainable Development in Mineral Exporting Economies. *Resources Policy* 19(1): 1429.

Bell, R. M. 1990. *Continuing Industrialisation, Climate Change and International Technology Transfer.* A report prepared in collaboration with the Re-source Policy Group, Oslo, Norway, and Science Policy Research Unit, University of Sussex.

Brown, R., and P. Daniel. 1991. Environmental Issues in Mining and Pet-roleum Contracts. IDS *Bulletin* 22 (4).

Coppel, N. 1992. *Worldwide Minerals and Metals Investment and the Environ-ment, 1980–1992.* Unpublished report for RTZ Corporation Plc.

Crouch, D. 1990. Personal communication.

Eggert, R. G. 1992. Exploration. In *Competitiveness in Metals: The Impact of Public Policy*, edited by Merton J. Peck, Hans H. Landsberg, and John E. Tilton. London: Mining Journal Books.

EPA (U.S. Environmental Protection Agency). 1986. Gregory Tailings, EPA, Denver. EPA *Region VIII Superfund Program Fact Sheet.* (July).

———. 1989. Smuggler Mountain, EPA, Denver. EPA *Region VIII Superfund Program Fact Sheet.* (March).

———. 1990. Silver Bow Creek Site, EPA, Montana. EPA *Region XVII Superfund Program Fact Sheet.* (May).

Johnson, C. J. 1990. Ranking Countries for Mineral Exploration. *Natural Resources Forum* 14 (3): 178–186.

Johnson, Stanley P., ed. 1993. *Agenda 21.* Book II of The Earth Summit: The United Nations Conference on Environment and Development. London/Dordrecht/Boston: Graham & Trotman Ltd./Martinus Nijhoff.

Jordan, R., and A. Warhurst. 1992. The Bolivian Mining Crisis. *Resources Policy* 18(1): 9–20.

Kelly, D. R. 1990. Personal communication.

Kennecott Corporation. 1992. Press release, March 11.

Kneese, A., and C. L. Schultze. 1978. *Pollution, Prices, and Public Policy.* Washington, D.C.: Brookings Institution.

Kopp, R. J., and V. K. Smith. 1989. Benefit Estimation Goes to Court: The Case of Natural Resource Damage Assessments. *Journal of Policy Analysis and Management* 8 (4): 593–612.

Milliman, S. R., and R. Prince. 1989. Firm Incentives to Promote Technological Change in Pollution Control. *Journal of Environmental Economics and Management* 17: 247–265.

Mining Journal. 1990. Inco Count Down Acid Rain. 23 February.

———. 1992. 30 October.

Mining Magazine. 1991. Scuddles Innovative Recruitment. (January): 9.

Nuñez, A. 1992. Heterogeneity of Production and Domestic Technological Capabilities in Mining and Mining Related Productive and Service Activities in Peru: Their Relevance for an Environmental Strategy. Research proposal. Lima, Peru.

O'Connor, David C. 1991. Changing Patterns of Mineral Supply and Demand. Draft discussion paper presented at the Organization for

Economic Co-operation and Development Centre's Mining and Environment Research Network Workshop No. 1, Wiston House, United Kingdom, April.

OECD (Organization for Economic Co-operation and Development). 1991. *Environmental Policy: How to Apply Economic Instruments*. Paris: OECD.

Office of Technology Assessment, U.S. Congress. 1988. *Copper Technology and Competitiveness*. Washington, D.C.: U.S. Government Printing Office.

Orr, L. 1976. Incentives for Innovation as the Basis of Effluent Charge Strategy. *American Economic Review* 56: 441–447.

Panayotou, T., Q. Leepowpanth, and D. Intarapravich. 1990. Mining, Environment, and Sustainable Land Use: Meeting the Challenge. Synthesis paper no. 2, 1990 TDRI (Thailand Development Research Institute) Year-End Conference, 8–9 December, at Jomtien.

Skea, J. 1994. Environmental Issues. Chapter in *Handbook of Industrial Innovation*, edited by M. Dodgson and R. Rothwell. United Kingdom: Edward Elgar.

U.S. Bureau of Mines. 1989. *Bauxite, Alumina, Aluminum Annual Report*. Washington, D.C.: U.S. Government Printing Office.

———. 1990. *Copper Annual Report*. Washington, D.C.: U.S. Government Printing Office.

———. 1991. *Copper*. Washington, D.C.: U.S. Government Printing Office.

Warhurst, A. 1990. *Employment and Environmental Implications of Metals Biotechnology*. World Employment Programme Research Working Paper, WEP 2-22/SP.207. Geneva: International Labour Organisation.

———. 1991a. Metals Biotechnology for Developing Countries and Case Studies from the Andean Group, Chile and Canada. *Resources Policy* 17(1): 54–68.

———. 1991b. Technology Transfer and the Development of China's Offshore Oil Industry. *World Development* 19 (8): 1055–1073.

———. 1992. *Mining and Environment Research Network Newsletter* no. 2 (April).

———. 1994. *Environmental Degradation from Mining and Mineral Processing in Developing Countries: Corporate Responses and National Policies*. Paris: OECD Development Centre.

Warhurst, A., and L. J. MacDonnell. 1992. Environmental Regulations. Chapter 2 in *Mining and the Environment: The Berlin Guidelines.* London: Mining Journal Books.

Womack, J., D. T. Jones, and D. Roos. 1990. *The Machine That Changed the World.* New York: Macmillan.

Printed and bound by CPI Group (UK) Ltd, Croydon, CR0 4YY

21/10/2024

01777049-0007